WHO ARE YOU?

Strengthening Personal Identity Management in Australia

S.J. Venning

Published in Australia in 2017 by Steven Venning

Email: svenning8@gmail.com

© Steven Venning 2017

National Library of Australia Cataloguing-in-Publication entry:

Author:	Venning, S.J. – author.
Title:	Who are you? : Strengthening personal identity management in Australia / S.J. Venning.
ISBN:	9780648050902 (paperback)
Subjects:	Identity theft – Australia – Prevention.
	Personal information management.
	Identification cards – Australia.

Disclaimer

The author has made every effort to ensure the accuracy of the information within this book was correct at the time of publication. The author does not assume and hereby disclaims any liability to any party for any loss, damage or disruption caused by errors or omissions, whether such errors or omissions result from accident, negligence, or any other cause.

Edited by Epiphany Editing & Publishing
Cover design and typesetting by Epiphany Editing & Publishing
Printed in Australia by SOS Media + Print

CONTENTS

PART FOUR – Personal Privacy and Data Security

PART FIVE – Enhancing the Core Identity Products

PART SIX – Overseas Systems for Identity Management

PART SEVEN – Future Trends in Identity Management

INTRODUCTION

Within modern Australian society, identity crime is the most prevalent form of crime. Since the 1980s, identity crime has been rising exponentially and today it is estimated to cost the Australian community over $2.2 billion per annum. Australian governments at federal and state/territory levels have recognised the severity of this crime and are committed to tackling the issue. Similarly, identity crime is increasing across most Western societies and many overseas governments are also actively attempting to tackle this escalating problem.

Identity crime is defined as the activities or offences in which a perpetrator uses a fabricated identity, manipulated identity, or a stolen or assumed identity to facilitate a crime. Identity crime often occurs through the use of stolen or fraudulent identity credentials. Criminals are able to access the community's identity credentials online, by email, through social media, by operating scams and through organisational data breaches.

Continuing to adopt a 'what we've always done' approach to identity management has failed to resolve this costly problem of safeguarding Australian community members' personal identity. Australian governments have started investing to improve identity management, but more work is necessary to

1

fill 'the gaps' that still exist in the government's model for improving identity management processes and systems.

Traditionally, within the Australian community, the primary form of personal identity document was the driver's licence – and this is still the case today. Customarily, if a person did not have a driver's licence they used their birth certificate as proof of their identity. Both driver's licences and birth certificates are issued by state and territory government bodies and, for this reason, identity management in Australia has been the responsibility and daily business of the states and territories.

In contrast, the federal government issues passports (which many Australians do not possess) and provides Australian citizenship certificates to new Australians. Likewise, the federal government issues a Medicare card which although held by a very high percentage of adult Australians is not considered to be a primary proof of identity document because it lacks sufficient personal information.

Australian state and territory transport authorities are responsible for issuing drivers' licences to qualified drivers in the community. The driver's licence contains the person's full name, date of birth, gender, home address and driving and card details. Importantly, it is one of the few secure government-issued documents that also incorporates an image of the person's face, as well as their signature. Approximately 90 per cent of Australia's adult population have a driver's licence that they usually carry with them.

The purpose of the driver's licence is to regulate drivers and their driving performance within the Australian community. This emphasis on being aware of the identity of the driver is to ensure penalties and other sanctions relating to poor driving performance can be allocated to the correct person. Transport authorities use an over-arching principle in identity management which is 'one driver, one licence'. This is designed to ensure that a person is known by their official name or identity and can only be in possession of one driver's licence in that name at any one time.

The role that state and territory transport authorities provide in terms of identity management in the community has been further acknowledged through the introduction of a non-driving identity card which is referred to by a number of names (including the adult age card) across Australia. The adult age card is issued as proof that a person is over 18 years of age and can legally enter a pub or club where alcohol is served. However, it incorporates the same personal identity processes and functions as a driver's licence and is therefore considered to be a secure government issued identity document. The adult age card is held by many non-drivers (such as the elderly) and is often used by them as proof of their identity.

Within this book a range of proposals are recommended that are designed to assist Australian governments to better protect the Australian community from identity crime through strengthening existing identity management policies, processes, systems and products. These proposals support the commitment that has already been made by federal and state/territory governments to better coordinate their efforts in identity management. The proposals also ensure the integrity of the community's privacy and data security provisions continue to be maintained.

In a move to strengthen identity management within Australian society, it is proposed that government agencies support the principle of 'one person one identity'. This initiative means that government agencies must ensure they know a person's official identity and only that identity is used to determine a person's eligibility for government services and benefits.

The proposals to strengthen identity management presented in this book are intended to quickly and effectively improve identity management in Australia. They are also designed to be very cost efficient for government to implement while, at the same time, having no (or only minimal) direct impact on the vast majority of Australians.

In essence, the proposals target improvements that Australian governments can implement to better protect the community and require no (or little) action or cost by members of the Australian community.

PART ONE

Overview of Identity Management

ONE

Background to Identity Management

Identity management is the cornerstone of government service delivery in modern Western societies. Providing the community with secure identities for its citizens ensures scarce resources are correctly allocated to those who are eligible. Increasingly, Western governments are focusing on identity management as a primary response to exponentially rising levels of identity crime (including identity theft) since the 1980s. Today, identity crime is the most common form of crime in many countries, including Australia.

Identity management practices have evolved over the years. Initially, it involved a person 'having' something to prove their identity. This was typically a physical document or token. With the advent and widespread adoption of personal computers in the 1980s and 1990s, identity management practices changed to something you 'have and know' to prove a person's identity. Consequently, having a document or token also required a PIN or a password, a secret or knowledge of someone's personal history. Today, with new biometric technologies available, identity management is shifting to 'something you are'. This means that a person's identity is now linked to a unique personal characteristic that signifies a person as a distinct individual.

Shifting identity management practices to 'something you are' means that, in the future, there could be much less reliance on the production of physical documents or knowledge of secrets for people to prove their identity. Many consider the shifting of identity management to using the unique characteristics of a person to be 'the utopia in identity management'. Operationally, it means organisations are not required (or have fewer requirements) to issue identity products to clients, yet they are still able to maintain a certified and secured client identity.

Since the 1980s, the growth in social security spending in most Western countries has risen steeply (and ever increasingly) towards a level which many leaders think will be unsustainable into the future. There are also fundamental issues for communities in dealing with social impacts due to some families having a multi-generational dependence on social welfare.

The burden on the community's social welfare system is increasing and will continue to do so as many Baby Boomers head towards retirement. This is putting increasing pressure on governments to reduce welfare spending and better target spending; an objective which may be achieved to some extent by reducing fraud costs within the welfare system. Improved identity management is a critical aspect in any governmental attempt to reduce fraud-related welfare payments.

Following on from the terrorist attacks on the United States on 11 September 2001, another factor that has highlighted the importance of strong identity management is border security. In most Western countries there is an increased priority on knowing who is entering a country, whether they should be allowed into the country, how well they are likely to assimilate within the existing community and the overall security risks that new immigrants present to a country.

The purpose of this book is to provide detailed proposals and an implementation strategy designed to quickly improve personal identity management in Australia and to better position the Australian community for the future. The proposals in this book may have some impact on the issue of border protection but, on the whole, they are more strongly focused on making significant improvements to Australia's domestic management of identity.

Traditionally, Australian governments have not focused strongly on identity management. The exception has been Australian transport authorities who issue driver licence products. State and territory transport authorities are the leaders in identity management in Australia and this has been the case for an extended period of time.

Since the mid-2000s, the federal government has started to increase its focus on identity management in response to rising levels of identity crime and the associated costs to government and the community. In December 2014 the Martin Place siege occurred in Sydney. As a result of the investigation into that siege, Australian governments have started to invest heavily in improvements designed to strengthen identity management.

The investigation into the Martin Place siege found that the criminal involved in the siege had interacted with a number of government agencies under a range of identities, aliases and titles. Part of the problem was a lack of rigour displayed by some government agencies to ascertain the criminal's official name. The report also identified a lack of coordination and sharing of client identity information between government agencies.

Led by the federal government, an improved focus on identity management has been underway since 2015, although some fundamental improvements to the current plan may still be necessary. Identity management actions implemented so far have better positioned government organisations to defend the community from increasing levels of identity crime. In essence, federal, state and territory government agencies have moved to nationally align their standards, processes and systems used to maintain identity management in Australia.

TWO

Identity Crime in Australia

The Australian Federal Police define identity crime as a generic term to describe the activities or offences in which a perpetrator uses a fabricated identity, a manipulated identity, a stolen or assumed identity to facilitate a crime.

At the time of writing this book, the most up-to-date report on identity crime in the Australian community is *Identity Crime and Misuse in Australia 2016* produced by the federal government's Department of Attorney General. This report states that Australia loses $2.2 billion annually to criminals through identity crime. It also claims that every year the community loses a further $390 million in responding to identity crime.

The report highlights that the federal government loses $98 million annually to fraud and Australian citizens lose approximately $660 million as a result of personal fraud. Police records show a further $1.3 billion is lost to fraudulent activities. Each year approximately 4 per cent of the Australian community incurs some type of financial loss as a result of identity crime.

The Australian Bureau of Statistics has stated the incidence of identity crime in Australia has increased significantly over the past few years. Obviously, systems and practices in the management of personal identity need to be enhanced in order to slow the increasing trend of identity crime, let alone to reverse it.

A 'do nothing' or continue with what 'we've always done' approach to identity management will not be able to successfully prevent the rising incidence of identity crime and the associated increasing costs to the community.

Most commonly, identity crime occurs with the use of stolen or fraudulent identity credentials. Criminals are accessing identity credentials online, by email, through social media, by operating scams and through organisational data breaches. Most information is being stolen from computer devices. However, organisational data breaches have also increased from 71 to 110 in a 12-month period from 2014 to 2015. An organisational data breach gives criminals access to potentially thousands of clients' personal information which may then be sold onto other interested parties.

The two most common identity documents used in Australia for identity crime are the driver's licence followed by Medicare cards. The Attorney General's report states it is possible for someone to buy a fake or stolen driver's licence for approximately $400. In contrast, the approximate cost to purchase a passport is $5,000. Interestingly, it is reported that in most cases where a person's identity has been sourced from their identity credentials being fraudulently used, the original identity document is still in the owner's possession (who is now the victim of identity crime). In most cases, the victim is unaware that a crime has occurred using their identity until they are informed after the event.

The Attorney General's report highlights that approximately 40 per cent of victims of identity crime do not report the crime to police. The reasons for this include the fact that some people think that nothing can be done after the crime has occurred; some do not report the crime because only a small amount of money was stolen or lost; and some are embarrassed that the crime has occurred. Likewise, organisations who are victims of quite significant fraud may not report the crime to police in case it becomes public knowledge which, in turn, can harm their reputation within the community.

National Identity Security Strategy (NISS)

As previously stated, identity crime has been the major crime committed in the Australian community for some time. There is growing concern about its prevalence in the community and also the speed with which it is increasing. Some sections of the Australian community may consider the federal government to have been 'asleep at the wheel' in relation to developing a strategy to tackle the rising incidence of identity crime in Australia. However, this opinion may be unjustified to some extent.

Traditionally, the key personal identity document in Australia is the driver's licence. The other main identity document that people use if they do not have a driver's licence or in conjunction with their driver's licence is a birth certificate. Both these products are issued by the Australian states and territories. The states and territories were (and still are) the main providers of identity management in Australia.

The federal government's involvement in identity management largely revolves around the issuing of passports. Traditionally, few Australians have held passports. It is only relatively recently that the Australian community as a whole have become more prevalent overseas travellers which has resulted in a much higher percentage of the community holding passports. When a person applies for a passport they are required to undergo a process to prove their identity. The two most used documents that are provided as proof of identity to obtain a passport are the driver's licence and birth certificate.

In addition, the federal government issues Australians with a Medicare card that permits people to have access to health services. Although it is widely held in the community, the Medicare card is not considered to be a primary identity document but it may be used as additional evidence to support an identity claim.

The Medicare card does not show a facial image or a person's signature. Moreover, it does not show the person's full name, date of birth or gender. In many cases there are multiple names of family members on the document; allowing a person to nominate which name belongs to them. The Medicare card cannot be used as an identity document in its own right. However, it is used to claim medical benefits and since it is often used without other supporting identity documents, this may be a factor in rising medical-related costs.

Concerned with rising levels of identity fraud, the federal government has started to adopt a leadership role in combating this issue. A national leadership role to coordinate and strengthen the nation's defence against identity crime has been what was missing.

In 2005, the Council of Australian Governments (COAG) agreed that protection of the person's identity was a key concern and a right of all Australians. COAG is a council that comprises the federal and all state and territory governments, including the prime minister and state and territory premiers. At that meeting, COAG also agreed to the development of a national identity security strategy designed to protect the identities of Australian citizens.

Importantly, at the 2005 COAG meeting the council also agreed to the development of a national Document Validation Service (DVS), as well as focusing on new and emerging biometric security systems that are often used to protect a person's identity from theft.

Following the 2005 COAG meeting, in 2007 the federal government as well as all state and territory governments signed a formal agreement for a National Identity Security Strategy (NISS). The signing of the NISS agreement acknowledged the need for Australian governments at the federal, state and territory levels to cooperate to protect the community from identity crime and the fraudulent use of fake or assumed identities as a *national priority*.

The NISS agreement was a ground-breaking development that kick-started the process of standardising identity management across all levels of Australian government – although in some ways it may be considered as a 'catch up' to what was already in place within Australian transport authorities.

The NISS agreement was a major step forward for Australia governments to recognise that Australia had a problem with identity crime that needed to be addressed.

A strong focus of the NISS has been on government-issued personal identity products; the proof of identity process that is used to obtain government products that may subsequently be used as proof of identity; physical security that is embedded in government-issued identity documents; and information exchange between government agencies across federal, state and territory governments and across jurisdictional borders. The NISS also highlighted the desire to use biometric technologies to strengthen the identity management process in government agencies.

The NISS agreement also proposed to establish a National Document Validation Service to allow validation of state and federal identity documents that people can use as proof of their identity.

Much of the work to be done under the NISS agreement was already in place with state and territory transport authorities through their driver licence business. The state and territory transport authorities are to be commended for having done much of the 'heavy lifting' in relation to protecting the community's identity prior to the introduction of the NISS agreement. Over the years, in the absence of national leadership, these agencies have implemented uniform standards, processes and document securities to strengthen Australia's identity management while providing a national approach and coordination in identity management.

Transport agencies were the first organisations to improve the security of driver licence documents. This made it more difficult and costly for criminals to produce high-quality fake drivers' licences. This was an effective strategy designed to price criminals out of the identity market.

It was then necessary for state and territories to improve their standards and processes around driver licences. This is because it was seen as easier and cheaper by criminals to fraudulently obtain a real driver's licence through the licence process then to make a high-quality fake licence.

Transport authorities had already created standards and processes for managing client identity, some of which were adopted by the NISS development team. But the NISS agreement and subsequent standards and processes were important because it provided the framework for federal, state and territory government agencies to align their practices and processes with a national best practice model in identity management.

In the 1990s, the federal government funded the creation of the Austroads organisation. This organisation provides a forum for coordination between the states and territories on transport related matters, including road building. This helps to provide national consistency in government operations across jurisdictional borders.

A subcommittee within Austroads contains state and territory transport authorities. This subcommittee standardised proof of identity requirements and processes that are used to obtain driver licence products across Australia – an initiative which occurred well before the creation of the NISS agreement.

The use of these requirements and processes were extended to adult age cards that each transport authority also provides to the community. The adult age cards are commonly used as proof of identity in the Australian community and are often used by non-drivers.

During the late 1990s, Australian transport authorities funded the establishment of a national service that allows transport authorities in all states and territories the ability to check a driver's licence or vehicle registration regardless of the state or territory that issued the product. Managed by Austroads, this service is also available to police across the nation.

Transport authorities have always recognised the importance of the driver's licence product to the Australian community's personal identity management. These authorities invested in strengthening the product and supporting processes which occurred across Australia in a reasonably consistent manner as the community's defence against identity crime.

Encapsulated in the NISS agreement was the recognition of the importance of establishing a national Document Validation Service to check identity documents. Transport authorities had already established this service but it was restricted to only checking their products (which were driver licence and vehicle registration related).

An area in which transport authorities were less successful was in making online validation of driver licence products through the national service available to other government agencies and the private sector. However, it should be noted that a number of transport authorities were legislatively restricted from providing wider access to driver licence information.

Some state and territory governments were concerned about possible litigation that may occur if a driver's licence is issued in error and then subsequently used by a criminal to defraud a person or organisation. The position transport authorities adopted was to focus on situations where the driver's licence was issued for the purpose of driving and road safety, and not as an identity product. However, in recent years this position is changing. The NISS agreement was a major turning point to this traditional position adopted by state and territory governments.

However, even after the NISS agreement was in place and development of personal identity standards was completed, there is little evidence of further NISS-related development. The Martin Place siege that occurred in December 2014 changed this and might be considered as a watershed moment for personal identity management in Australia.

The 2015 investigation into the Martin Place siege highlighted a general lack of rigour by some government agencies in complying with the NISS standards for identity management. An outcome of the investigation was that federal, state and territory governments reaffirmed their commitment to cooperate and improve identity management. They have since invested – with some urgency – in developing other aspects of the 2007 NISS agreement.

THREE

Personal Identity Products in the Australian Community

The Australian Federal Government estimates there are currently over 50 million proof of identity documents that are being used within the Australian community. These documents are issued by over 20 federal and state government organisations. This count does not include documents that are used to support proof of identity claims that have been issued by private sector organisations.

Traditionally (and still the case today), government agencies require the provision of physical documents as proof of a person's identity. A primary document is defined as a secure government-issued document that contains a facial image or photograph of the person. For an organisation to establish a person's identity, usually a primary and a secondary identity document are required.

Documents being provided as proof of identity to obtain a driver's licence or adult age card are often grouped into three types. The first type of document is proof that an identity or person exists. A birth certificate or immigration document provides this proof. The second type of document is one that proves the identity exists in the community. A Medicare card, a credit card or other relatively high-value document could fulfil this requirement. The third type is a lesser (but still ideally government-issued) document that contains the person's address such as an electricity bill, a rates notice or a rental agreement. This information proves a person is a permanent resident of a state/territory and establishes their official address.

The documents described above are of particular importance if a person does not have a primary government-issued identity document that includes their facial image and they are applying to obtain such a primary government identity document. Providing an organisation with a volume of different documents can also support a person's ownership claim to an identity

The most common products used in the Australian community to support identity management are listed below.

Debit / Credit Card – Debit and credit cards are some of the most prevalent cards in the Australian community. It is estimated that there are over 44 million debit cards in active use within the community and a further 16 million credit cards.

Debit and credit cards are issued by Australian financial institutions such as banks. Clients are legally required to prove their identity before a financial institution can establish an account. This is done in accordance with the federal government's *Anti-Money Laundering and Counter-Terrorism Financing Act 2006*. This legislation requires a person to provide a number of specified identity documents to prove their identity. This legislation created the 100 point identity check that is still used by financial institutions today.

Like the proof of identity standard that transport authorities use when people apply for a driver's licence, the 100 point identity check involves a list of documents that a person could use to prove their identity. A point value was given to each type of identity document that could be provided. To successfully prove ownership of an identity, a person must provide enough documentation to ensure the combined point value of the documents is equal to 100 points.

Typically, a debit or credit card does not contain an image of the cardholder and is not used as proof of identity. It can be used by government agencies as evidence of the use of an identity in the community. The use of debit and credit cards in identity management is as secondary support to a person's ownership claim to an identity.

Medicare Card – The Medicare card is issued by the federal government for the purpose of obtaining medical services. The card does not contain an image of the person and is not used by government agencies or financial institutions as a primary identity card. Although it is a government-issued card, like debit and credit cards it is used in identity management to support a person's claim to an identity.

Loyalty Cards – These are typically low-value cards that are issued by private organisations specifically for the purpose of obtaining services from those organisations. It is rare that these cards contain a client's image and it is unknown what vigour has occurred with regards to checking the person's identity prior to issuing the loyalty card. Loyalty cards are not accepted as proof of identity documents and have no value in the identity management process by government agencies or financial institutions.

Passports – Passports are highly secure primary personal identity products. The passport product is considered the highest value proof of personal identity product in the community. A passport contains the person's full name, date of birth, gender, facial image and signature and product details. The Australian passport document is very secure against tampering and complies with international passport standards.

There are two problems with the passport product being used within the Australian community. The first issue is that a large number of Australians do not hold a passport. The second issue is the fact that passports are rarely carried domestically. The design of the passport meets international requirements for an international travel document, but it does not easily fit into wallets and purses carried by Australian citizens. It is often used in the initial or first-time identity process when a person first registers their identity with an organisation. When not travelling, it might be used as a special occasion identity document.

Driver Licences – Driver licences are secure primary personal identity products. A driver's licence is held by approximately 90 per cent of the adult Australian population. All Australian drivers' licences are a nationally consistent product. They are plastic, credit card sized products that contain the person's full name, date of birth, gender, facial image and signature, address, driving and card details.

Most wallets and purses are designed to easily display a person's driver's licence on the inside of the item. Consequently, the driver's licence is the most carried form of personal identity document in the community.

Many government and private sector systems and processes are geared towards the use of the driver's licence to prove a person's identity. Some private sector organisations have aligned their records systems to record driver licence numbers and details. Other types of identity documents such as passports may be refused as proof of identity by some organisations because their systems are not designed to be able to record them.

Birth Certificate, Citizenship, Immigration Certificate –
These documents usually have a number of security features
and so are considered reasonably secure. Although they usually
do not contain a person's facial image, they are considered to
be primary identity documents.

These documents show the official name of a person
which is evidence that the identity exists within the community. Typically, these types of documents are not carried
but are provided when a person first registers their identity
with an organisation, such as a government department or a
financial institution. The documents are always used with a
number of other supporting documents that used in combination prove a person's identity.

FOUR

Biometric Technology in Identity Management

C onsidered by many people to be the path to the future, biometric technology is a process whereby electronic systems recognise people's unique characteristics. When used with a database, this technology can match a person's biometric information with a person's identity. Biometric technology analyses the unique features or characteristics of a biometric, such as a person's face or fingerprint or voice. Biometric systems work where a person's identity is first registered in a database and linked to the person's particular biometric such as a digital facial image.

A biometric identity management system could be used to verify a person's identity by collecting a copy of the person's biometric such as digital facial image and matching it with a copy of the facial image that is stored in a database. Once the match is confirmed, the system provides the person's full identity information. It can also be used to identify a person by comparing a person's digital facial image with all other facial images in a database to find a match, which then again confirms the person's identity information from the database.

Biometric technology can reduce the community's reliance on carrying documents as proof of their identity and can improve and speed up access to approved services.

There are a range of biometric technologies in use today. Some of these technologies have been around for a long time while others have only recently evolved into being high-capacity and robust technological systems. Within the Australia community, popular demand for using biometric technologies has really only occurred in the last 10 years in response to government concerns about identity security and reducing fraud, particularly in relation to government benefits.

In the past two years, there has been a strong Australian Federal Government led push to improve identity management, including expanding the use of biometric technologies as a defence to rising terrorism and identity crime rates. Internationally, biometric technology is increasingly being used to improve border control.

Biometric technologies are being widely used by governments across the world. The two most common forms of biometric technologies are facial recognition and fingerprinting. In Australia, the New South Wales, Victoria and Queensland state transport authorities use facial recognition technology with digital facial images for issuing drivers' licences.

Australian's Department of Foreign Affairs and Trade uses facial recognition and fingerprint recognition for some Australian passports. Some countries such as the United States require fingerprints as part of their border protection policy. As a result, people can expect that in the future fingerprinting may become part of the standard passport application process.

International groups such as the Middle East's Gulf Cooperative Council (GCC) (which is a group of six countries within the Gulf region) allow a person to use an approved identity card containing smartcard technology and their fingerprints to enter and exit their countries. In this case, a passport is not required. These six countries use biometric fingerprint technology to match the result of a fingerprint scan to an algorithm of the fingerprint that is securely held in the smart chip on the smartcard identity card to confirm someone is able to enter or leave their country. A more detailed explanation of this passport control system is provided later in this book using an example from the United Arab Emirates.

The most widespread use of biometrics occurs in India through their government-sponsored Aadhaar Biometric ID Program. This system uses fingerprint recognition and has a database of 1.1 billion people or identities that it uses to verify a person's identity. In 2016 the Indian service was processing approximately 15 million identity verification transactions per day. Most biometric technologies are very robust. The Indian program is used to support the provision of government services, including the payment of benefits to Indian nationals. The program was established to cater for the high proportion of the Indian population who do not have identity documents and in an effort to reduce corruption. It was estimated that approximately 40 per cent of government benefits paid to Indian citizens were previously lost due to corruption.

The concept of biometric technology to automatically identify or verify a person's identity is considered by many to be the future for managing personal identity. Conceptually, it means that the reliance for people to carry physical documents to prove personal identity is reduced or not required once a person's biometrics and identity are enrolled in a biometric

27

system. This approach matches the Indian service described previously.

While biometric technology is available and is currently very good, the technology poses a number of risks involving data security and personal privacy. At present, governments are the main users of biometric technology which usually operate within secure government systems. However, the risks to the community are greatly increased if this technology becomes used extensively by private sector organisations.

Private sector organisations may not have the same level of controls and safeguards in place as government agencies. Moreover, there may be discrepancies in terms of the level of financial investment that an organisation assigns to safeguarding data and privacy. Some private organisations may not place the same priority on data security and client privacy as governments do. This was evident with the recent Yahoo data breach were personal details on approximately 500 million Yahoo users were stolen and then became available for sale to criminals.

In regard to the privacy concern, facial recognition systems in particular have a huge capacity to track people. Over the past few years, American (USA) law enforcement agencies have invested in facial recognition systems with the goal of improving biometric system performance to better provide one-to-many facial matching.

Facial recognition technology is reliant on a high-quality image of a person's face to allow a match to occur with a stored copy of a person's face. Stored facial images are usually collected in 'perfect' environments such as an office where the subject and the lighting are controlled. In fact, most facial recognition systems confirm to a camera operator that the client's facial image quality is adequate immediately after the image is collected and before a client moves away from the camera.

However, American (USA) law enforcement agencies are seeking to improve the performance of facial recognition technology in real-time street environments for one-to-many facial image matching. American agencies have tested setting a digital camera at the entrance to a football stadium that is connected to a facial recognition system and a database of images. As spectators enter the stadium, the camera collects their facial image and matches it against a database of potential terrorists and criminals. The aim of this process is to ensure that the biometric system identifies any terrorists and/or criminals who have entered the football stadium.

Privacy groups are concerned that this approach means that the camera is recording *everyone's image* and, potentially, a government would be able to record every person who walked through the football stadium entrance without their knowledge. Theoretically, this could be achieved by linking the football stadium camera to the state transport authority driver's licence image database.

Biometric technology provides governments with a very easy and effective tool to track the movement of a large number of citizens throughout the community. The result from the test described above was that further work was necessary in terms of the quality of images being captured in a live (street) environment (although the biometric system was able to manage the checking process of the volume of images of people walking through the stadium entrance).

As mentioned in the Indian government example, modern biometric systems have a huge capacity to manage personal identity and provide identity/verification services. However, an issue not often discussed by technologists is the process involved in re-establishing a person's identity (including their biometric) after it has been stolen by criminals.

In Chapter 15, I discuss the process of recovering a stolen identity which involves governments reissuing personal identity documents with different client and document references to someone who has had their identity stolen. However, re-establishing a stolen identity that is linked to a person's biometric that has also been stolen is conceptually and operationally a very difficult task.

Some people have focused on this problem as being a negative aspect of biometric systems. In practice, this should not be an issue where a person's biometric is being collected live by an organisation as part of a business transaction. A scenario where the issue may arise is in cases where a biometric is being collected by an off-site system for use in a transaction. For example, in cases where a person attends a government office to conduct a transaction and a fingerprint is collected from them, it would be very difficult for a criminal to substitute someone else's fingerprint. In this scenario, the transaction is being fully managed by a government system. However, if a transaction occurs online and the person is using their computer to interact with a government system, it may be possible to use someone else's fingerprint or other biometric.

While unlikely, it is still considered to be an unresolved issue with biometric technologies. The issue becomes more problematic if transacting becomes more reliant on a biometric that is being received from a remote non-government site or system.

To date, there is no information available that a person's biometric has ever been stolen from a government system and used in a fraudulent transaction.

In the case of transport authorities, most of these organisations have the person's identity information and facial image stored in two different systems. In theory, someone could break into the image database and steal a copy of a person's facial image. However, there is usually no client identity information stored with the image, only an image reference number. The criminal would need to break into a second government system and use the image reference number to identify who the person is – a very unlikely scenario.

The widespread use of biometric technology by governments reduces the risks of not initially accurately identifying an individual. Once a person with a fake identity is recorded on a government system with their biometric, the person cannot be then registered again under their real identity. This may discourage criminals from using fake names to obtain benefits; in particular, where biometric checking occurs nationally.

Brief descriptions of some of the more common biometric technologies that are available and in use today are listed below.

Iris Scan – This process involves scanning the iris of a person's eye using a digital camera. The person places their eye approximately 20 cm from a camera to allow a copy of their iris to be recorded. It is a very accurate technology with approximately 200 data points being collected from the iris for comparison with stored copies of the person's iris scan. In comparison, a fingerprint only has 60 or 70 data points available for use when matching this biometric.

Iris scanning is often used in highly secure but low-volume applications. It is an intrusive form of technology and some people may resist this technology so it may not be widely used in the community.

Fingerprinting – Fingerprinting is the oldest and most widely used biometric technology. Traditionally, fingerprinting has been used by police to identify criminals. Today, fingerprints are increasingly used by governments and private sector applications such as personal computers for access control. A fingerprint is collected for use in a biometric system by placing the finger on an electronic scanning plate. The fingerprinting technology is easy, accurate and fast to use but within the Australian community may still be associated with criminals.

Voice Record – Voice print technology is most often used to support access to services that are provided by telephone. To be registered, a person normally reads a script that is recorded and registered against their identity details. Of all the technologies listed in this book, it is the only one that involves the use of telephones and may be the most unreliable of the biometric technologies. Failures may be caused by external factors such as background noise, poor telephone connection or audibility issues.

Facial Image – Facial recognition technology maps the features of a person's face from a digital facial image of the person. Facial image technology is the most widely used non–intrusive biometric technology and is used almost exclusively by governments. In Australia, facial image technology is used by some state transport authorities that collect client facial images to produce driver licences. It is also used by the federal government in relation to Australian passports.

The technology works by comparing a new facial image with all other images in an image database and highlights where matches occur. A match means that the person is potentially already recorded in the image database, possibly under another name or identity. A manual check can then be undertaken of the two matched images to confirm the person is not recorded under a different identity. This technology is a key defence against the creation of fake identities or the theft of an identity.

FIVE

Current Identity Management Practices in Australia

The driver's licence has become the de facto Australian identity card. While there are a variety of different identity documents that may be used in the Australian community most are issued once a person has confirmed their identity, which often occurs by showing their driver's licence as proof of identity.

Most Australians use their driver's licence as the primary proof of identity document. Part of the reason for this is that it is one of the few secure government-issued documents containing the person's facial image and personal identity details. All Australian drivers' licences are credit card sized products that easily fit into wallets and purses. As a result, the driver's licence is nearly always carried by cardholders.

Another factor that supports the widespread use of the driver's licence as a proof of identity document is the product's high penetration within the Australian community. Approximately 90 per cent of adult Australians hold a driver's licence. Adult Australians are those who are over 18 years of age. Most states allow learner driving from 16 years of age. When a person applies to a state or territory transport authority to start learning to drive they are required to meet a nationally consistent proof of identity standard. This process is managed by state and territory transport authorities, who issue driver licence products.

State and territory transport authorities also issue an adult age card that is typically sought by non-drivers, such as the elderly. This product is known by a variety of names across Australia such as the Proof of Age card. However, the purpose of this card is to prove a person is over 18 years of age and able to legally enter pubs and clubs that serve alcohol.

To obtain an adult age card a person must meet the same proof of identity standards and processes as if they were applying for a driver's licence. The adult age card has the same securities and personal identity information as the driver's licence. It is widely used in the community as a government-issued identity document and accepted as proof of identity the same as driver's licence.

Some years ago, the personal identity framework used by state and territory transport authorities to confirm the identity of a person for issuing a driver's licence or adult age card was standardised across Australia. The driver licence product and information that appears on the driver licence product or adult age card is very consistent across all Australian states and territories.

Approximately 20 years ago, Australian states and territories funded the establishment of a central national service called the National Exchange of Vehicle and Driver Information System (NEVDIS). This system manages the exchange of driver licence and vehicle registration information between the states and territories. NEVDIS is used on a daily basis by transport authority officers who are involved with driver licensing and vehicle registration services, as well as police from across the nation. The system makes it possible for a transport authority officer or police officer to quickly check a driver's licence or vehicle registration (including ownership and driving / vehicle history) regardless of which state or territory issued the product.

Of all the driver licence products issued in Australia, the Queensland driver licence is considered to be the most advanced and secure driver's licence in the country. The Queensland driver's licence contains a contact smart chip or minicomputer that is embedded within the plastic driver's licence card that can be used to store data. This makes it very hard to fraudulently reproduce a Queensland driver's licence.

Information stored on the card can only be read by inserting the driver's licence into a smartcard reader that is connected to a computer. The data that resides on a smart chip can also be stored in an encrypted format that further secures the data from unauthorised access. This means that if data on the smart chip does not decrypt properly then the product may be a fake.

When a person applies for a driver's licence within Australia part of the process involves the relevant state or territory transport authority collecting a digital image of the client's face. The person's digital facial image is printed onto their driver's licence when it is issued. Furthermore, all Australian transport authorities retain a copy of the client's digital facial image in a state/territory based image database.

In this way, state and territory transport authorities currently hold facial images of all Australian drivers. The adult age card looks similar to a driver's licence and also contains a facial image of the cardholder which is also stored by transport authorities. Transport authorities store the image mainly to allow duplicate products to be produced if the original is lost.

Transport authorities in Queensland, New South Wales and Victoria use facial recognition technology within their image databases to reduce the potential for identity crime to occur using a fraudulently obtained driver's licence or adult age card.

When a person renews their driver's licence or adult age card, a new facial image is collected. The facial recognition technology does an automatic comparison between the new facial image that has been collected and the previous facial image stored in the database to ensure they match and it is the same person. When a person is first issued with a driver's licence or adult age card, the technology compares their facial image with all other images in an image database of drivers and adult age card holders to ensure the person is not already recorded with a different identity.

As a result of the very high level of contact that Australian transport authorities have with the community, they maintain very large service delivery networks that provide face-to-face contact with clients. In most cases, access to driver licensing services is available in-person from every large and medium-sized city and town across Australia.

Typically, transport authorities also provide remote service delivery of driver licences, such as allowing clients to renew or replace their drivers' licences via an online or tele-phone service delivery mode. However, when someone first enrols or registers as a new client they are required to visit a transport authority office in-person. This is necessary for the client to prove their identity to government staff by providing a range of proof of identity documents.

Successful completion of the proof of identity process establishes the person as a client of a transport authority and a fully-registered entity in the authority's database. At this time, the client is recorded with a customer reference number within the authority's database. In some states (such as Queensland), the customer reference number that is first allocated to a client after they are registered with a transport authority becomes the client's lifetime driver's licence and client reference number.

To support driver licence business activities, transport authorities manage extensive telecommunication and database systems. Transport authorities also provide nationally-consistent policies, processes and card production equipment to support the personal identity and driver licence business. These authorities also employ officers who are highly experienced in identity management to actively manage the nation's personal identity business on a daily basis.

PART TWO

Proposals to Strengthen Identity Management in Australia

SIX

A National Strategy for Identity Management

This book proposes that the best strategy to strengthen identity management in Australia is to continue to support a national approach to the current identity management processes and products that are already in place. An advocated proposal routinely raised is to introduce an Australian identity card that would be available to the community to support better identity management.

Ideally, any proposed identity management strategies should have no direct impact on the vast majority of Australians. In addition, they should be relatively easy and cost efficient for government organisations to implement. Furthermore, it is intended that the proposals discussed in this book should not raise significant concerns about client privacy and data security issues as they are designed to comply with existing standards and norms already in place to manage privacy and data security within the Australian community.

A key to any proposed strategy to strengthen identity management in Australia is to leverage additional benefit from existing investments in infrastructure, policies, systems and processes, products and expertise that already exists across the two tiers (federal and state/territory) of government.

Objectives of the Proposed Identity Management Strategy

The proposals for a national identity management strategy have the following objectives:

- *To strengthen identity management in Australia and better protect the community from identity crime.* The major aim should be to provide a high level of confidence within the community that a person is who they say they are. This objective also needs to include recommendations that make it more difficult to obtain identity documents in another person's name.

- *To reduce the incidence and associated costs involved in a person being able to fraudulently obtain government services.* This objective will ensure that governments can have a greater level of confidence that benefits and services are only being provided to eligible people.

- *For there to be no (or only minimal) inconvenience to the vast majority of Australian citizens.* For most Australians, the identity management proposals recommended in this book should have no impact, no cost and not require any action on their part.

- *To implement the proposals quickly and at low cost to government.* The proposed strategy is to use existing infrastructure, standards, processes and norms already in place in the Australian community.

- *To reduce privacy and security concerns within the community about the proposed identity management strategy.* It is important that individuals and groups are aware of how their privacy and data will be managed under these proposals. The intent is that there is no (or negligible) increase in community concern about an individual's privacy or the security of a person's personal data.

SEVEN

Key Principles of Identity Management

The fundamental principle of identity management discussed in this book is that for any individual who interacts with government organisations there is only one official identity. In this way, this principal of identity management states: 'one person, one identity'.

Even though a person may be known by many names within the community, in any dealings with government only their official name or identity is to be used. For example, John William Citizen who is 25 years old may be known in the community as Harry who is 22 years old. Within the community, this practice is completely okay. However, government agencies would only use the person's official identity (i.e. John William Citizen – 25 years old) when determining what services and benefits they are eligible for.

For some time, state and territory transport authorities have had a similar principle in place which underpins their

44

management of driver identities. This principle states: 'one driver, one licence'. As part of the driver licensing process, clients are expected to provide a number of identity documents and personal information to satisfy a particular authority as to their official identity. Their official identity is then registered by the authority and driver licence products are issued in that official name.

This strong interest by transport authorities in establishing a person's official identity is for the purpose of managing the driver licensing process (including the road safety aspect). It ensures that the relevant transport authority can be confident that penalties and sanctions that relate to poor driving behaviour are being allocated to the right person. This practice has been a common approach in states and territories for an extended period.

Similarly, the same processes and personal identity requirements apply to the adult age card where only the person's official identity or name is recorded and used by transport authorities to issue the adult age card.

In each state and territory in Australia, the Registrar of Births, Deaths and Marriages is the main (but only one of two) holder of Australian citizen's official identities. A person's official identity is recorded on their birth certificate.

A person's official identity may change with the recording of a change of name through marriage or through the legal name change process. The Registrar of Births, Deaths and Marriages manages all changes to a person's official identity. Transport authorities are reliant on documents issued by the Register of Births Deaths and Marriages when determining a person's official name.

The other holder of the community's official identities is the federal government's Department of Immigration and Border Protection. This department manages the Australian citizenship process whereby immigrants to Australia formally become Australian citizens. The citizenship certificate functions in the same manner as a birth certificate.

If a person (who was originally an overseas citizen) becomes a permanent resident of a state or territory but does not become an Australian citizen, they are entitled to use their overseas passport as their primary identity document when applying for government services and products.

EIGHT

Strengthening Identity Management Strategies in Action

This chapter outlines a range of proposals for action by Australian governments that are designed to strengthen identity management in Australia.

Proof of Identity Documents

One of the major proposals to strengthen identity management within the Australian community is for government agencies to only accept three core personal identity documents as primary proof of a person's identity.

These three core identity documents include:

1. an Australian driver licence
2. an adult age card
3. an Australian identity card (which is yet to be developed).

This proposal means that a client must produce a driver's licence or adult age card (or the new Australian identity card) as the main form of documentation that proves their identity to obtain a government service. Australian citizens should be aware that to prove their identity to government organisations, provision of one of the three core (or primary) identity documents may be required potentially with other supporting secondary documents.

Currently, there is no proposal to remove any existing identity products from the Australian market. Rather, there is a recommended strategy to concentrate core or primary identity onto the three core proof of identity products mentioned above. Any other identity products that are used within the Australian community should be considered to be secondary identity products. If another identity document is used without one of the above three identity documents, an organisation should not accept that a person has legitimately proved their identity.

For a person to establish their identity to a government organisation, it is proposed that only one of the three core identity documents is necessary. Once a client has officially established their identity by using one of these three core identity products, an organisation can register the person as a client and issue the person with the organisation's own identity product or document if they wish. The client could then use the organisation's identity product to identify themselves to that organisation in the future.

The driver's license and adult age card products should continue to display a person's personal identity information as well as card information. These products should continue to contain the same features and securities and be nationally consistent. The proposed Australian identity card should contain the same personal and card information and the same features and securities as available on Australian driver licenses. This includes digital facial images and signatures.

All facial images that are collected for an Australian driver's license or adult age card or Australian identity card should be checked using facial recognition technology. This is to confirm that a client is not known by another identity, or is not attempting to steal another person's identity through the issue of the above identity products.

The three identity products above should all contain smartcard technology. The smartcard should contain at least basic personal identity and card issuing information. This information should be stored in an encrypted format. Adding smartcard technology and data encryption to the three identity products is intended to make the products very difficult to fraudulently reproduce or alter.

Space should be available on the smart chip of each of the three identity products to store agreed government (federal, state/territory and local) data. Cardholders should also have the option to choose to store state/territory approved, non-government applications and data on their smart chip products.

Links should be established between transport authorities and all state and territory Offices of Births, Deaths and Marriages (OBDM). This is to allow the online checking of certificates issued from any Australian OBDM. Furthermore the OBDM should inform transport authorities when a person, who holds any of the three above identity documents, changes their name. Similarly a link should be provided to transport authorities to allow them to confirm the validity of an Australian citizenship certificate.

Validating Proof of Identity Documents

A change in governmental policy will strengthen the proof of identity process. When a person provides one of the above three documents as proof of their identity it should be implied that the government has approval to check the person's identity and associate document details of the provided document.

It is proposed that the federal government engages the state and territory transport authorities and request that they enhance their national information exchange service (which is known as the National Exchange of Vehicle and Driver Information System (NEVDIS)) to provide information on the projected Australian identity card and adult age card in the same way as they are currently using NEVDIS support for driver licences.

The NEVDIS service is now linked to the federal government's new Document Validation Service (DVS). But this DVS link provides only basic validation of driver licence products to public and private sector organisations.

Checking a Person's Identity

In order to check a person's identity using a driver's licence, adult age card or the proposed Australian identity card, an organisation needs to link to the federal government's Document Validation Service (DVS). An operator who is checking the client's identity should manually enter the client's reference number that appears on the card product. The DVS should then retrieve the following client and card information from the appropriate storage service: client facial image, client reference number, client name, client date of birth, gender, card expiry date and card number.

It is intended that an operator who is checking a person's identity is 'in possession' of the person's identity document in order to be able to enter the client reference number from the identity document. Furthermore, it is proposed that there would be no ability for an operator to randomly view client identity information from the DVS.

In time, the smart chip on the three core identity products could be used to automate the checking process via the DVS when it is inserted into a smartcard reader. At that stage, manual interaction with the DVS should no longer be available. Using the smart chip is to speed up the identity checking process and reduce the ability for an operator to randomly view client information from the DVS.

It should not be possible for an operator who is not with a state or territory transport authority to alter or delete information that is displayed on NEVDIS or the DVS. Only approved transport authority officers would be able to access this data which is a current process via NEVDIS.

Enhancements to the NEVDIS and the DVS

Enhancements are proposed to the NEVDIS and the DVS to improve information exchange between the states and territories and other non-transport authority organisations. For instance, currently only a driver's name, date of birth and driver's licence number is available to the Document Validation Service (DVS) from the NEVDIS. The enhancement should ensure that the client's image, as well as card information, is provided on the three core identity documents so that this information is available to the DVS.

A further proposed enhancement of NEVDIS is the development of a separate facial image database. This database would be designed to store a national copy of all driver licence, adult age card and Australian identity card facial images. Also, within this image database should be facial recognition technology to facilitate image checking.

In essence, this enhancement will form the basis of a national image database capable of performing national facial image checks. The image database should contain client facial images but not contain client identity information such as a person's name or client reference number.

NINE

Introducing the Australian Identity Card

It is proposed that the Australian Federal Government should be the owner of the planned Australian identity card. The federal government should determine who is eligible for the Australian identity card, the validity of an individual's application for the card and any fees associated with it. The Australian identity card should be available to anyone who meets the government's eligibility criteria.

However, the operational management/administration of the Australian identity card should be the responsibility of the state and territory governments. Management of the Australian identity card should closely mirror the current management practices for identity and identity card products used by state/territory organisations for driver's licences and adult age cards.

In this way, the state and territory transport authorities would become the service provider for the Australian identity card on behalf of the federal government. Transport authorities would continue to use the existing range of documents used by clients to prove their identity when first applying for a driver's licence or adult age card to also be used when applying for an Australian identity card.

This proposal also means that there is no recommendation to make the Australian identity card compulsory for Australian citizens. Obtaining an Australian identity card from a state or territory transport authority should be an option open to all Australian citizens.

At this point in time, the Australian identity card does not exist. In most cases, clients of government and other services use their driver's licence or adult age card as proof of their identity and the majority of Australians are expected to continue using these existing products to prove their identity. As a result, there should be no change in the method that most people in the Australian community currently use to prove their identity.

Currently, all Australian transport authorities issue a plastic PVC-type driver's licence card that contains a facial image of the client and their personal and driving details. In addition, Queensland's transport authority produces a PVC-type driver's licence that also contains a smart chip embedded into the card. The smart chip is a minicomputer that can hold data and be read using a smartcard reader linked to a computer.

Outsourcing the operational management of the Australian identity card to state and territory transport authorities will permit the use of existing infrastructure, policies, processes and equipment already in place to produce and manage driver licence and adult age card products. The outsourcing proposal will avoid the need for the federal government to duplicate costs in order to establish processes and systems (since these are already in place). It also engages transport authorities who are already the main providers of personal identity within the Australian community.

It is proposed that the Australian identity card would contain the same (or similar) personal identity and card information that currently appears on Australian drivers' licences. However, it is also envisaged that the Australian identity card would contain a smart chip (as is currently contained within the Queensland driver's licence product).

Ensuring that the Australian identity card is a smartcard product means that it will be highly resistant to fraudulent reproduction or alteration. It also means the card will have the potential to store data; a feature that may streamline future business transacting between clients and the government (and possibly also with private sector organisations).

Production of an Australian Identity Card

It is proposed that all applications for the Australian identity card could be electronically forwarded to the smartcard supplier for the Queensland Department of Transport and Main Roads (Queensland Transport) in order for them to operate as the centralised provider of the Australian identity card. Currently, this private supplier provides the smart-chipped Queensland driver's licence. The intention is that the Australian identity card will contain the same smart chip and overall card product (including card securities and features) as the Queensland driver's licence and be produced centrally by Queensland.

Using the existing contractor would facilitate the rapid introduction of the new Australian identity card at minimal cost to government in comparison to a full development of a new procedure (including an acquisition process that may potentially be necessary to engage a new provider). If a new provider has to be selected, more time and cost would be required for the supply and installation of the necessary equipment and services that are already present with the Queensland card provider.

Consistency Standards for Smartcard Products

The Queensland driver's licence and the Australian identity card should contain the same agreed information in the same encrypted format. It is proposed that arrangements are made with the Queensland government to allocate space on the Queensland driver licence smart chip for federal and local government services (and also for future data).

All state and territory governments should be encouraged to transition their existing driver's licence and adult age card to become smartcard products. These identity products should all contain the same federal government information on the smart chip.

Service Delivery Network for Australian Identity Cards

To fast track improvements to Australia's identity management, it is proposed to allow clients who are interested in obtaining an Australian identity card to apply at any state or territory transport authority office across the nation. Transport authorities should process applications and identify clients using the same policies, processes and standards that they currently use to determine a person's identity for a driver's licence. Client facial images should be forwarded to NEVDIS for national facial image checking and then the image and other personal information should be sent to Queensland Transport for production of the Australian identity card.

The Australian Identity Card and Personal Identity Information

It is suggested that the following information could appear on the Australian identity card:

Card Front

- Client facial image
- Client signature image
- Client reference number
- Client name
- Client date of birth
- Gender
- Card expiry date

Card Back

- Issuing authority
- Statement of card purpose
- Card reference number

Smart Chip

- Card reference number
- Issuing authority
- Card expiry date

The intention is that client identity and card information that is available from NEVDIS is the same as that which appears on the driver licence or adult age card or Australian identity card. There should be no access to additional client information from NEVDIS.

Facial Images

All state and territory transport authorities collect a digital facial image of their clients when they issue them with a driver's licence or adult age card. These authorities retain a copy of the client's digital facial image on their local image database.

In the case of the states of Queensland, New South Wales and Victoria, they also use facial recognition technology within their image databases. Facial recognition technology checks a new image against other images in the database to reduce the likelihood of a client creating a fake identity or stealing another person's identity through the driver licence or adult age card processes. This automated checking process supports the 'one person, one official identity' principle.

As part of the Australian identity card application process, transport authorities should also collect a digital copy of the client's facial image and other personal information, such as their signature, to be held in their existing local database systems. Where a client for an Australian identity card already holds a driver's licence or adult age card, the client's existing facial image and signature (which already appears on their driver's licence or adult age card) should be used to produce the Australian identity card.

Transport authorities who collect client facial images for the production of driver licences, adult age cards or Australian identity cards should forward all facial images to a national image database within the NEVDIS service for national facial recognition checking.

TEN

Government Responses to Strengthen Identity Management in Australia

The following agreements will be required within federal and state/territory government organisations to enable the proposals to strengthen identity management in Australia described above to proceed.

Federal Government

- The federal government should agree to the development of an Australian identity card and they should be the owner of the card.

- The federal government should engage state and territory transport authorities to provide operational management for the Australian identity card as formal service providers.

- The federal government should ensure that the Australian identity card information is available on the Document Validation Service (DVS) and that it matches the identity information provided for a driver's licence.

Government Agencies

- Government agencies should agree that the only core or primary proof of identity documents that will be accepted to obtain services are: a driver's licence, an adult age card or an Australian identity card.
- Government agencies should invest in smart card readers to allow information on the three smartcard identity products to be read.

State and Territory Transport Authorities

- State and territory transport authorities should agree to use their existing infrastructure, processes, equipment and expertise in identity management to also provide the Australian identity card.
- Centralised production of the Australian identity card and national identity checking by transport authorities should occur.
- Australian state and territory transport authorities should agree to accept applications for the new Australian identity card and transmit the applications to Queensland Transport for card production.
- States and territories should agree to an update of the NEVDIS service to include an independent national facial image database that contains facial recognition technology.
- State and territories should agree to provide the national facial image database with facial images that

are collected for the purpose of driver licensing, adult age card and Australian identity card for checking.

- State and territory transport authorities should agree to update NEVDIS to provide client facial images to the DVS, as well as other card and identity related data.

- State and territory transport authorities should investigate the exchange of client identity information with the Registrars of Births, Deaths and Marriages and the Department of Immigration and Border Protection. Investigation should further occur into the information exchange occurring via the NEVDIS system.

Queensland Department of Transport and Main Roads

- The Queensland Department of Transport and Main Roads (Queensland Transport) should agree to arrange the production of the Australian identity card that is to be highly resistant to fraud.

- Queensland Transport should arrange to add encrypted data to the smart chip in the Queensland driver's licence, adult age card and the Australian identity card in consultation with the federal government.

- Queensland Transport should agree to provide space on the Queensland driver's licence and adult age card smart chip products for federal government information.

- Queensland Transport should agree to accept digital facial images and client identity data from other state and territory transport authorities in order to produce Australian identity cards.

Registrars of Births, Deaths and Marriages

- Registrars of Births, Deaths and Marriages in each state and territory should agree to establish online links to state and territory transport authorities – possibly through the NEVDIS service. This is to achieve improved national business efficiency by utilising an existing service if possible.

- Registrars should agree to provide transport authorities with online validation of personal identity documents that they have issued.

- Registrars should agree to provide transport authorities with change of name and death information on clients who hold an Australian driver's licence, adult age card or an Australian identity card.

- Transport authorities should be able to validate birth certificates, marriage certificates and legal change of name certificates that clients provide as proof of their official identity when applying or changing government-issued identity products.

Department of Immigration and Border Protection

- The Department of Immigration and Border Protection should agree to provide state and territory transport authorities with an online link to allow these authorities to validate an Australian Citizenship Certificate. This may also be achieved by an information exchange through the NEVDIS system.

PART THREE

Core Identity Products in Australia

ELEVEN

Verifying the Official Identity of Australian Citizens

A key principle discussed throughout this book is the focus on an individual's official identity when transacting with government organisations. In Australia, people's official identities are verified through one of two processes:

1. being born in Australia or an Australian territory, or
2. via the Australian citizenship process.

Births, Deaths and Marriages

Management of the community's official identities of people born in Australia is a state and territory function. Each state and territory in Australia has its own independent organisation – Office of Births, Deaths and Marriages (OBDM) – who are responsible for maintaining the record of official identities of all people born in their state or territory.

In order to obtain a driver's licence or adult age card, a combination of primary and secondary identity documents are used by a person to prove their identity. The birth certificate that is issued by the Office of Births, Deaths and Marriages is the most commonly used primary identity document to obtain a driver's licence.

The proposal to provide online access for state and territory transport authorities to the Office of Births, Deaths and Marriages is to facilitate the validation of birth certificates and other documents issued by the office. It is also proposed that transport authorities have the ability to place an identifier against a person's official record in the Office of Births, Deaths and Marriages which indicates the person has been issued with a driver's licence, adult age card or Australian identity card.

The proposal for online access to records held in OBDM is a national requirement. For example, a person may be born in the Northern Territory but later becomes a resident of Western Australia (WA) and applies for a WA driver's licence. For this reason, cross-jurisdictional access to records held in the Office of Births, Deaths and Marriages must be made available on a national basis.

As previously discussed in Chapter 8, transport authorities already use a cross-jurisdictional system to exchange driver's licence information – the National Exchange of Vehicle and Driver System (NEVDIS). It is possible that cost savings could be achieved if an enhancement to NEVDIS occurs by providing the same functionality of exchange of information between transport authorities and Offices of Births, Deaths and Marriages.

Introducing a process whereby the transport authority puts an indicator against a person's name in the Register of Births, Deaths and Marriages will allow the OBDM to advise the relevant transport authority if the person changes their name or becomes deceased.

Having access to advice from the OBDM means that the state and territory transport authorities may be less reliant on the provision of documents from people who attend offices to have their name changed on their driver's licence or adult age card or Australian identity card. In the case of a death notice, it also allows the client's record to be updated whereby no further activity on the record can be expected. For instance, this action may reduce the likelihood of some drivers attempting to transfer traffic offences that are captured by speed cameras to the records of deceased drivers.

Australian Citizenship Records

Similarly, it is proposed that a link be established between transport authorities to the federal government's Department of Immigration and Border Protection to allow citizenship certificates that are presented as proof of identity to obtain a driver's licence, adult age card or Australian identity card to be validated. The purpose of this validation process is to confirm the document's accuracy and the person's official identity.

TWELVE

Aligning Australian Identity Products

The goal of aligning the Australian identity card with the driver's licence and adult age card products is to reduce community concern about a national identity card. Most people within the community are already familiar with the process of attending transport authority offices, proving their identity and being photographed to receive a driver's licence or adult age card. Consequently, using the same personal identity policies and processes as those that support the driver's licence product may create a familiarity or a perception of a routine process within the community for the Australian identity card.

While the Australian identity card would be a federal government product, outsourcing operational management to the states and territories may provide people with a greater level of comfort with regards to the new product. In contrast, an attempt to provide the new Australian identity card solely through the federal government may create higher levels of mistrust and uneasiness in some segments of the community.

During recent years, within some sections of the Australian community, there has been a perception of a general decline in the integrity of politicians, including federal politicians. There have also been concerns about poor performance by some federal government departments and a lack of empathy towards clients. For example, between 2014 and 2017 there have been ongoing examples of public rorting of travel expenses by some Australian government politicians which may have eroded community perceptions of the integrity of politicians in general.

There have also been accusations reported by the Australian media that the federal government's Department of Human Services has issued erroneous debt claims against clients in relation to Social Security overpayments. Some erroneous claims have highlighted incompetence by the Department of Human Services and have included a perceived unwillingness by the department to resolve issues with their clients when the department has been found to be at fault. These examples may all serve to promote community resistance to a fully federal government managed national identity card initiative.

It is not proposed to make the process of obtaining an Australian identity card mandatory. Instead, if a person who holds a driver's licence also wishes to obtain an Australian identity card, they should be able to achieve this objective without the need to be re-photographed or to attend a government office. The stored driver licence image and client details should be used to produce the new Australian identity product.

By allowing the community to continue to use the driver's licence and adult age card as proof of identity, it is expected that demand for the Australian identity card is likely to be low. However, this product may appeal to the non-driving members of the community such as the elderly.

Currently, the adult non-driving segment of the community primarily uses the adult age card as a government-issued proof of identity document. The adult age card is issued by each state and territory as proof that a person is over 18 years of age and can legally enter pubs and clubs that serve alcohol. This card is inappropriate for use by the elderly who may instead become the main holders of an Australian identity card.

Since there is no need for the majority of Australian citizens to obtain the Australian identity card because they can continue to use their existing driver's licence or adult age card, it is expected that concerns about a national identity card will be greatly reduced. For the majority of Australians, there would be no change to the current practice of using their driver's licence as proof of identity. Consequently, there will be no additional costs involved for driver licence and adult age card holders unless they wish to obtain the new Australian identity card.

The federal government would be responsible for determining who is eligible for the new Australian identity card. Logically speaking, it would be straightforward to permit anyone over 17 years of age to be eligible to apply for the new identity card. This age limit generally aligns with the age limit for issuing driver licences.

While most Australian states allow a person who is 16 years old to commence learning to drive, most states do not allow a person to drive solo until they have reached the age of 17 years. In the case of Victoria, this ruling is 18 years of age. When a person applies for a learner's permit which is required to learn to drive, their identity is established and recorded with the relevant state or territory transport authority. However, a facial image of the client usually is not collected until the first driver's licence for solo driving is issued which is usually 17 years of age.

However, it should be considered that access to the Australian identity card could be provided to clients who are under 17 years of age. The benefit of providing a strong identity document to clients who are under 17 years of age means they will have the community's standard identity document earlier in their life. This may allow these clients to benefit from a higher level of government protection of their identity. Typically, individuals who are under 17 years of age are not involved in fraud or other criminal activities where they could potentially want an identity document issued in a fake name. For this reason, there is less risk involved in establishing a person's identity at an early age.

Nevertheless, one issue associated with providing an Australian identity card to clients who are under 17 years of age is that, throughout their teenage developmental years, they can have significant facial changes which would require their identity document to be updated more frequently than the usual ten-year period. Updating the document more frequently may incur additional costs to the client and government.

There are overseas examples where identity cards are issued to young people but these must be updated more regularly than for older people to ensure the client's facial image on the identity document continues to match the person's face.

In the case of driver licensing, when a person is first issued with a solo licence it is usually a probationary licence at 17 years of age. The person is again photographed when they transition to an open driver's licence which normally occurs three years after the first probationary licence – meaning the client is at least in their early twenties. A client's facial image does not change significantly once they are in their early twenties. This allows transport authorities to produce drivers' licences which closely match the client's face while only having to update the client's facial image every 10 years.

An important proposal discussed in this book is to adopt facial recognition technology to strengthen the identity management process. In theory, a person may be successful in establishing a fake identity through the driver licence process. However, with the facial recognition technology in place, it would then be impossible to establish their real identity with a transport authority.

Informing the community about the use of facial recognition technology may mean that criminals will be less likely attempt to establish a fake identity through the driver's licence, adult age card or Australian identity card products. This is due to difficulties they may encounter when they then attempt to establish their real identity.

Linking Official Identity Documents

The federal government's new Document Validation Service (DVS) allows other organisations to check on a person's identity by using driver licence products and is a very good first step in strengthening identity management practices. However, as previously outlined in Chapter 8, there are two key questions that are important in identity management. One is that a person's identity formally exists. Another is that the person in question owns the identity. The current DVS arrangement only provides information that the identity exists.

At present, the DVS does not provide a link between the official identity document and an individual. This objective could be achieved by state and territory transport authorities also providing the DVS with the client's facial image that is recorded on their driver's licence (or other document) and the associated document details. Providing the facial image to the

DVS is a major step to improving identity management and closing an existing gap in the national identity management framework.

Another concern with the current DVS is that it does not validate adult age cards which are often used by non-drivers as personal identity documents.

Establishing a closer relationship between the states/ territories and the federal government may also support improved data transfer where illegal activities have occurred. For instance, information about aliases that have been used for fraudulent purposes could be passed between the states/ territories and the federal government to facilitate improvements to government records and reduce risks to government.

The state and territories authorities have access to this data but it is not shared with federal government agencies or other organisations that link through the DVS. This information is currently shared between Australian police services so an enhancement to NEVDIS and the DVS could ensure this information is provided to other government and possibly private organisations.

Official Identity Management within the Australian Community

A large proportion of Australians have been known by at least two official identities during their lifetime. Typically, this occurs for women when they make the change from their maiden to a married name. The change of name process is managed at the state and territory level between the Registrar of Births Deaths and Marriages and transport authorities who issue a new driver's licence with the new name.

It should be noted that most state and territory transport authorities have legislation in place that requires a person to notify the authority when the person changes their name or their address. In the case of Queensland, this notification period is only 14 days. As a result, the driver's licence product is very accurate in terms of providing up-to-date client information.

Obviously, the value of an identity document is much greater where there is confidence that the document is up-to-date. The same requirement to notify the issuing state or territory authority of a change of name or address should be applied to the adult age card and the Australian identity card.

Improved interaction with the federal government through the DVS may facilitate faster updating of federal government records with a person's new identity. The update process reduces the ability for a person to obtain government services under both their previous and their current official names or identities.

Some characteristics of the driver's licence and adult age card products and businesses that should be the same for the Australian identity card are listed below:

- the products are secure and government issued
- the person must prove their identity during an in-person visit
- they contain the person's full official name, date of birth and gender
- they include a digital facial image of the client and a digital image of the client's signature
- all of the products are nationally consistent
- Australian drivers' licences are consistent with international standards for driver licences
- identity requirements for a driver's licence and adult age card are nationally consistent
- the driver's licence and adult age card are supported by extensive service delivery networks

THIRTEEN

Australia's Austroads Group

Austroads is a federally-funded agency that creates a forum for coordination between the federal, state and territory governments with a particular focus on the Australian transport industry and related activities. Within Austroads, a subcommittee exists that is made up of representatives from each state and territory transport authority, as well as representatives from the New Zealand transport authority. The subcommittee focuses on Australian national consistency of driver licensing and vehicle services.

Over the years, members of the Austroad's subcommittee have been instrumental in closely aligning driver licence and vehicle registration policies and business practices across Australia. For instance, this subcommittee established the nationally agreed standard that is used to formally identify a client who applies for a driver's licence. Much of this standard was adopted by the federal government in *National Identity Proofing Guidelines* of 2016 that occurred under the NISS agreement.

While there is no obligation to do so, many of the standards, processes and protocols adopted in Australia are also used by New Zealand authorities in the management of drivers and vehicles.

The *National Identity Proofing Guidelines* is also used within a number of private organisations to improve consistency, standards and processes in identifying clients who apply for their identity products.

NEVDIS and the DVS

In the late 1990s, the National Exchange of Vehicle and Driver Information System (NEVDIS) was established by transport authorities with operational management provided by Austroads. The service does not hold client data but, instead, provides a link to the issuers of driver licences and vehicle registration services (i.e. the transport authorities). NEVDIS receives a request for information (such as driver licence information) and forwards the request to the relevant transport authority. The relevant transport authority provides the information to NEVDIS which is then available to the requesting authority or police. NEVDIS is a 24-hour seven day week service.

The federal government's Document Validation Service (DVS) was established in 2009. While it was agreed for the DVS to be established in 2006 under the NISS agreement, its expansion only commenced after the Martin Place siege that occurred in Sydney in December 2014. The DVS operates in the same way as NEVDIS but it assists a higher number of clients who link to the service. The service provides available information on a large range of identity documents – not just drivers' licences – that are used within the community.

When the DVS receives a request to verify driver licence information, it forwards the request to NEVDIS who manages the retrieval of information from the relevant transport authority. This information is then provided back to the DVS. Like NEVDIS, the DVS does not store client information.

As mentioned previously in Chapter 8, the NEVDIS and DVS both require enhancement to provide facial images of cardholders (driver licence, adult age card and Australian identity card) and other relevant card information.

NEVDIS and a National Facial Image Database

As broadly described above, it is proposed that the states and territories establish a national facial image database that is managed as a subsystem of NEVDIS. This image database could be potentially managed through the Austroads structure and should contain facial images of all people who hold drivers' licences, adult age cards and Australian identity cards. The image database should also contain facial recognition technology that facilitates comparisons of new images that are collected for the three core identity products against those already stored in the database.

A transport jurisdiction that is responsible for collecting a new digital facial image when a person applies for one of the three core identity products could send the facial image to the national database with an image reference number, the reference number of the client's previous image and an indication of the organisation who collected the image.

The image database should provide a facial recognition check on the new image against the previous image (in cases where a previous image exists) to confirm it is the same person.

WHO ARE YOU? – Strengthening Personal Identity Management in Australia

It should then provide a one-to-many image matching process that encompasses all images in the database (a national check) to confirm a person is not attempting to create a fake identity or to steal another person's identity.

If the image check confirms the image is okay, a response should be provided to the associated transport authority which should then send the image and client details to the Queensland Department of Transport and Main Roads for central card production and distribution to the client.

If the image check highlights a problem, then advice should be sent to the transport authority that is interacting with the client for investigation and resolution.

Identity Safeguarding and Investigation

Each state and territory transport authority should be responsible for investigating a person who has applied for one of the three core identity products and is suspected of attempting to establish a fake identity or to steal another person's identity.

Establishing a national image database may assist in identifying criminals who are involved in fraud and are operating across multiple states and territories. Investigating these criminals would require coordination between the states and territories, as well as the relevant police departments. While there is already a high degree of cooperation between the states and territory transport authorities, some transport authorities may need to develop additional expertise in investigation and in the preparation of evidence that may be used to support criminal prosecutions.

FOURTEEN

The Australian Identity Card and Federal Government

The federal government should be the owner of the Australian identity card and should be responsible for determining the eligibility criteria, card design, the card validity period and any fees associated with the provision of the Australian identity card. It should also determine how the card is provided to the community and what type of card validation is available to support the personal identity process.

These considerations should occur in close consultation with state and territory transport authorities since they are the traditional leaders of Australia's domestic identity business.

It is proposed that state and territory transport authorities be the service providers for the Australian identity card. To effectively engage the states and territories in this proposal, a reimbursement to them as service providers should be considered.

It is recommended that state and territory transport authorities would provide:

1. Identity management services
2. Operational policy for identity management services
3. Australian identity card service delivery and card management
4. Australian identity card details to the DVS for product validation, as well as assisting in verification of the cardholders' identities.

Smart Cards and Identity Products

It is recommended that all transport jurisdictions update their driver's licence and adult age card so that they are smartcard-enabled products; the same system as is currently provided in Queensland. As well as strengthening the integrity of the product, the addition of smartcard technology to drivers' licences supports the possible introduction of automatic product validation, while, at the same time, strengthening data security and client privacy.

Smartcard technology is also ideally suited to providing clients with a secure electronic identity that may be used to support better identity management and safer online transacting. This proposal extends the value of these products and better positions the community for future online interaction. It mirrors the initiative currently being rolled out by countries in the European Union that will be discussed later in this book.

Utilising smartcard technology means that access to client identity information through the Document Validation System (DVS) may potentially occur automatically when the smartcard driver's licence (or other identity product) is inserted into a smartcard reader. In this way, an operator who is attempting to verify a person's identity does not need to manually enter data into the DVS or other relevant system. It is important to note that the term 'automatically identifying a person' can only occur after the smartcard identity product is inserted into a smartcard reader.

Removing the requirement to manually enter data to access client information from the DVS ensures a DVS operator is 'in physical possession' of the driver's licence or other smart card identity product. Using smartcards in the identity process streamlines and speeds up the validation and verification processes.

The ability to automatically identify a person using the electronic identity that is stored on the smart chip of an identity product may also be extended to other non-government services.

In this book it is proposed that, in the initial introductory phase, the Australian identity card would be produced nationally by Queensland Transport. In this way, a person would interact with their local state transport authority which would verify their identity, collect a digital copy of the person's facial image and other details necessary for the production of an Australian identity card.

Following a national check of the person's facial image, this information would then be forwarded to Queensland Transport for card production. It is recommended that those state and territory transport authorities who do not currently produce smart-chipped identity products should also consider using this same process to centrally produce their state / territory smartcard driver's licence and adult age cards.

Engaging Queensland's smartcard contractor to produce smartcard driver's licence and adult age card may fast track the national introduction of smartcard drivers' licences and adult age cards into other state and territory communities. Using the Queensland model to introduce the smartcard products into other states and territories may be considered a short-term solution while they make arrangements to produce their smartcard products locally.

A critical consideration in the expansion of smartcard technology across Australia is the interoperability issue associated with the smart chip. It is more likely that computer hardware and software 'read and write' issues with the smart chip will be avoided if the same contractor provides all the smartcard products to the Australian market. There are also expected to be volume-based cost reductions achieved for states and territories from this approach to sourcing the relevant smartcard technologies from the one provider.

Smartcard Applications

Previously, in Part Two of this book, a number of examples were provided which highlighted the ways in which a smart chip contained in the driver's licence, adult age card and Australian identity card could be used. Generally speaking, the smart chip permits a person's official identity to be quickly and automatically retrieved from back office systems which, in turn, may streamline general transacting processes.

Inserting the smart chip product into a reader could also quickly record a client's card reference number without providing the client's actual identity (which could then be sourced at a later time if necessary). A good example of how this technology could be used at pubs and clubs to support Queensland government considerations of using identity as a means of reducing alcohol-fuelled violence will be discussed in Chapter 21.

Another example of the use of smartcards may be to support the government's electoral voting process. All

Australian citizens have a mandatory requirement to cast a vote in government elections. For an individual, there are two functions that occur when they are involved in the voting process: one is that their identity is recorded and the second is when they make their selection about their preferred candidate.

Voters could potentially be automatically identified by inserting their smartcard into a reader at a voting poll. In this case, there is an electronic national roll of people who are eligible to vote. Inserting the smart card would identify the voter and their record would be automatically marked as having attended a polling office. This method would speed the process of identifying clients who attend polling booths and may improve the accuracy and integrity of the voting process. For instance, it would prevent the incidence of a person voting more than once in the same election and also for people who are recorded as deceased from voting.

Adding data to the smart chip in a strongly encrypted format makes the data very secure from unauthorised access. The smartcard technology is very suitable for holding identity information that may support more secure online transacting, particularly with government organisations. However, clients need to have access to smartcard readers before they could utilise the security and flexibility that smartcards can provide.

Smartcard Readers in Action

Of course, using electronic identities that are stored on smartcards to support identity validation and for online transacting cannot occur without the rollout of smartcard readers. On the surface, smartcard readers may be considered to be a 'chicken and egg' problem. Do we provide the community with an electronic identity and then provide them with a smartcard reader or is it better to first provide the smartcard reader and then the electronic identity? Providing the community with smartcard readers may be a costly and a long-term problem for government and hence is not recommended.

One cost-effective approach may be to provide the Australian community with an electronic identity and then encourage people to purchase their own smartcard reader if they think it would be of value for them in better protecting their online identity. For example, a smartcard reader can be purchased for under AUD $20 and attached to a personal computer without specialised support required. Some people however may not want to purchase a smartcard reader because they do not currently transact online.

Security Benefits of Smartcard Technology

Today, access to most online accounts occurs through a user ID and password. The user has to submit something 'they know' (i.e. user ID and password) as proof of their identity to access their online accounts. An overseas criminal who has acquired a person's user ID and password would have the same access to a person's accounts and may conduct transactions (such as fund transfers) from the person's accounts.

By changing the 'online' proof of identity process to: (1) something 'a person knows' (i.e. user ID and password) and (2) something 'a person has' (i.e. an electronic identity stored on a smart card that is inserted into a smartcard reader) makes it much less likely for criminals to be able to access a person's online account. In this example, the criminal must also acquire the smartcard product to gain access to a person's online account, which is unlikely.

I am sure that many people would consider the purchase of a $20 smartcard reader that protects them from online criminals and potentially from financial loss to be a good investment!

If the Australian community is provided with secure electronic identities it may support other changes to public and private systems and services, such as the process of accessing online bank accounts which, in turn, may better protect community members. Another example is the potential use of electronic identities on enhanced social media services. The anonymity of some interactions could continue although a person's identity could be sourced, if needed. In this situation, knowing that the identity of a party who may be involved in an online discussion can be sourced, if needed, may result in the modification of certain behaviours or activities (such as a paedophile conducting an online discussion with a young child).

As mentioned above, a parent may consider the $20 purchase of a smartcard reader to be a good investment in protecting their child from online predators.

Creating a large population of people with secure electronic identities may incentivise public and private sector organisations to invest in enhancing existing services (as well as creating new services) that utilise the electronic identity. Establishing a large population of potential online users within the Australian community increases the likelihood that a critical mass of service users will quickly come on-board to support a new online service. This may speed up the return on investment associated with the new development and may make investment in new online initiatives by Australian organisations more attractive.

Rollout of Electronic Identities

An operational benefit of attaching an electronic identity to the three core identity documents (particularly the driver's license) ensures 100% of adult Australians are automatically provided with an electronic identity whenever an individual renews their identity product. Most Australian drivers' licenses have to be renewed every five years. This guarantees a five-year rollout period of electronic identities across the community. It also means the smart chip technology can be updated every five years as part of the standard replacement card process.

Importantly, after only two years, a significant percentage of the Australian population (particularly young people) would have an electronic identity that has been provided with one of the three core identity products. Under this proposal, after only two years, the federal government will have enabled a large proportion of the Australian community to conduct highly secure online transactions. Strategically, under this proposal, the federal government is extending its role in protecting the Australian community into the online environment by providing an electronic identity to citizens as part of the three primary identity documents.

FIFTEEN

The Role of Driver Licences in Identity Management

In Australia a person may have their driver's licence suspended or cancelled due to poor driving behaviour. In the case of a driving suspension, in some states and territories, the licence is retained by the person and they can continue using it as proof of identity. In situations where a driver's licence is cancelled, the licence product must be surrendered to the relevant state or territory transport authority or police. This means the individual no longer has access to this primary personal identity product.

Currently, a client in this situation could apply for an adult age card which provides the same personal identity function as the driver's licence. According to the proposals presented in this book, the person would have the option to choose to apply for either an adult age card or the new Australian identity card.

The flip-side to the current situation is where a secure personal identity has been added to an Australian driver's licence. Over the years there has been some discussion internationally about the possibility of revamping a driver's licence product so it becomes a smartcard application that resides on a national identity card. To date, this idea has never been developed.

A major problem associated with this concept is the ability to easily read driver's licence information. In a technologically advanced country such as Australia this concept might work. However, difficulties may arise in situations where a person is using their driver's licence (which contains a smartcard application) but there is no ability to read the smart chip, such as when they are overseas. Overseas police and others from the transport industry (such as car rental companies) must be able to quickly determine whether or not a person is licenced to drive a vehicle.

Even though it would be operationally very difficult, with today's advances in technology and telecommunications in Australia it may be possible to develop this concept of a driver's licence being an electronic application on another product. However, this could be done while acknowledging that a different solution would be needed for Australian overseas travellers who wish to drive.

Replacement Driver's Licences

Australian driver licences are typically issued for five-year periods. However, some licences (such as probationary licences) are issued for between one and three years. In the case of some elderly drivers, their driver's licence may be issued for a period of less than five years. In contrast, Victorian drivers have the option to obtain a ten-year licence. In the case of Queensland, the driver's licence is issued for a maximum period of five years. The smart chip in the Queensland driver's licence also has a guaranteed life of five years.

In terms of client facial images that appear on driver licences – nationally these images are updated every 10 years unless the person's face changes so much that there is no longer a match between their current image and their driver's licence facial image. In this case, a new facial image of the client is collected before the ten-year period. Similarly, a client's signature appears on the driver licence product. The signature image is also updated every 10 years unless the client requests an update before this timeframe has elapsed.

Since licensing authorities retain the client's facial image and signature and other personal details it allows replacement licences to be provided via remote service channels such as telephone or internet service delivery.

In the case of a driver's licence that is lost or stolen, an individual would be able to easily apply for a duplicate licence. In this situation an identical product could be produced to the one that was lost or stolen. The only difference between the two licences would be a different card reference number (the unique client reference number would remain the same).

It is proposed that the same processes and policies that are currently in place for drivers' licences should also be used to renew / replace the Australian identity card product when necessary.

Re-Establishing a Person's Stolen Identity

Australian transport authorities also provide an important but little used service which is to re-establish a person's identity after it has been stolen. Up until the last few years, this was an ad hoc service; however, as a direct result of the Martin Place siege, a formal court-managed process was developed to re-establish a person's identity after it has been stolen. The process allows a victim of identity theft to obtain a Commonwealth Victim's Certificate that is issued by a state or territory magistrate as evidence that their identity has been stolen and to support a formal process to re-establish their official identity.

This court process allows a range of organisations that provide identity documents (such as transport authorities) to establish a nationally-consistent process for re-establishing a person's identity. The establishment of this court process also recognises the prevalence of this crime in the Australian community and the devastating impact that the theft of a person's identity can have on them and their financial future.

SIXTEEN

Cost Effectiveness of Changes to Identity Management

A key component of the proposals put forward in this book involves focusing proof of identity requirements on the three primary or core identity products. The premise underlying this concept is to permit government to quarantine its future expenditure on the three products at the expense of other secondary identity documents. This strong focus on the three primary identity documents by government is to ensure they remain high-quality and high-integrity products with effective supporting processes.

Cost Effective Solutions to Identity Management

It is recommended that governments should be active and invest in the latest technologies to support the three core identity products. A desired outcome of this investment is to ensure the Australian community has the highest level of confidence in the identity products and their associated processes.

Consequently, government should not be expected to invest heavily in the security features that could be included in secondary identity documents. This approach may avoid or reduce the costs associated with acquiring increasingly complex security features and technologies that may be placed on secondary identity documents. Overall, focusing on the three primary identity documents at the expense of other secondary documents may lead to cost savings for government and private sector organisations.

However, an obvious investment that could prove to be cost effective in supporting secondary documents would be to provide online validation of secondary identity documents by document-issuing organisations, potentially through the government's Document Validation Service. Online validation with the organisations who issued the three primary identity documents *must be a priority*; however, online validation to support secondary documents would only be considered as desirable.

Ensuring a high degree of confidence in the integrity of the three core identity products and their associated processes could allow other public and private sector organisations to reduce their identity management costs as a result of only requiring an individual to use one of the three primary identity documents to prove their identity.

Moreover, the direct and indirect costs of identity management to an organisation would be transferred to the state and territory transport authorities. For instance, an organisation who currently issues their clients with a card/identity product (such as a golf club membership card) may instead request the client use one of their core identity products to identify themselves. As a result, the golf club would not need to continue to issue a membership card to its clients.

If the golf club needs to have a record of a member's attendance they would be able to do so by automatically recording the member's card reference details from the smart chip once it is inserted into a smartcard reader. Another more complex approach may be where the client agrees for an organisation, such as a fuel company, to place the organisation's smartcard application onto the client's smartcard identity product. In this case, the relevant transport authority would need to first agree to allow access to the smart chip to place the application.

In the example outlined above, clients may find that there is less necessity for them to carry a range of cards in their wallet or purse and instead use one of their core identity documents (such as their driver's license) more frequently.

Engaging the states and territories to produce the new Australian identity card avoids significant cost duplication for the federal government. It also speeds up implementation by using existing state and territory infrastructure, as well as existing policies, processes, equipment and card stationery. Importantly, this approach would also engage state and territory government officers who are already located in most medium and large towns and cities in Australia. These officers are already experts in identity management, customer service and the overseeing of identity products. The benefit of engaging these officers in this proposal should not be under-estimated.

For those people who already hold a driver's licences and wish to apply for an Australian identity card, this process could be done remotely without the need (and associated inconvenience) for them to visit to a government office in person. This, in turn, reduces service delivery costs for clients and government.

With this proposal in operation, it may be that some clients who hold a driver's licence and an Australian identity card will never need to attend a government office to apply for or renew their Australian identity card since it can always be produced using the data that was initially collected for their driver's licence.

The proposal to establish a central facial image database with facial recognition technology and communication links may be expensive; however, it avoids the need for expenditure and duplication by individual states/territories who have not already invested in the facial recognition technology. It may also provide the opportunity to avoid ongoing expenditure associated with individual state and territory image databases by instead relying on a national service.

There may also be the opportunity to recommission an existing image database used by one of the transport authorities as the national database. If this is possible, it may avoid some of the costs associated with establishing a new database, including hardware and software costs.

For the Australian Federal Government to establish itself as the community's main personal identity provider would be a new and very costly business. It is also more likely that some segments of the community may resist the proposal due to concerns about data security and privacy concerns. The outlay to the federal government of duplicating existing infrastructure, policies, processes and equipment is also expected to be costly and time-consuming if contracting state and territory transport authorities to be the Australian identity card service providers is not the chosen model.

As mentioned previously, this proposal should be balanced by the knowledge that demand for the Australian identity card is expected to be low because citizens may continue to use their driver's licence and adult age cards as their main proof of identity documents.

Implementing the Proposed Changes to Identity Management

Operationally, this proposed policy change would not be too difficult to achieve. Government agencies are already accepting drivers' licences as proof of identity so there would be no change in the processes currently used by the majority of government clients. Under this proposal, it is expected that most Australians would not purchase an Australian identity card but, instead, continue to use their driver licence as proof of identity. However, the new identity product may prove to be popular with non-drivers.

If accepted, implementation of this proposal is expected to occur quickly; potentially within one year if the states/territories agree to support the proposal and use their existing identity management infrastructure. If the federal government decides to establish and operate this new identity management business, an extended timeframe would be likely. Plus, the downside of the federal government managing this business would be the need to duplicate infrastructure, equipment and skills that already exists in the states and territories.

State and territory transport authorities are expected to support the proposal which may be seen as an extension of their existing business and an opportunity to generate additional revenue as service providers of the Australian identity card. A range of consultations between all levels of government would need to occur with regards to the Australia identity card proposal. This may occur more quickly if the Australian identity card product uses policies, processes and protocols that align closely to those currently used for driver's licences.

There is a need for a policy change in government agencies whereby only the three core identity products can be used as proof of identity. The policy change would mean that other forms of identity would not be accepted by government organisations as primary identity documents.

Potentially, enhancements to NEVDIS and DVS may occur within a one year timeframe which would allow additional information to be provided to DVS users from transport authorities. The main improvement to the current DVS would be to provide clients' facial images.

PART FOUR

Personal Privacy and Data Security

SEVENTEEN

Maintaining Client Privacy

According to policies discussed previously in this book, government organisations should have the right to check any of the core identity products that have been provided as proof of a client's identity. Checking a client's identity should involve viewing the client's identity information that is available from transport authorities. This objective may be achieved by accessing the information through the DVS.

The ability to access client information via the DVS is to require an operator (whether employed by the public or private sector) to be 'in possession' of the identity product that is being used by the client as their proof of identity document. The operator should enter the unique client reference number that appears on the card. The reference number is then matched and the client identity and card information is to be displayed from the DVS.

All the information available from the DVS record would also be visible on the card. There would be no additional information provided through the DVS. An operator who views the information from the DVS would have no access to alter, delete or copy the information. Only state and territory transport authorities would have the ability to do so.

The client information displayed on the Australian identity card should be the same as that which appears on the DVS for a driver's licence. This alignment of the three identity products is to reduce any additional privacy issues associated with the proposal to provide an Australian identity card. In the case of a driver's licence, the card displays additional information such as the client's address plus the vehicle classes the person is permitted to drive. This additional driving information is not considered personal identity information and should not be available from the DVS.

Some Australian transport authorities use a smart number as the client's unique reference number. A smart number is developed using an algorithm and usually the last digit is a check digit. When a number that contains a check digit is entered into a computer system, the computer software decodes the number and matches the check digit. This means the operator has entered the number correctly. In this case it also makes it very hard to access client information by randomly entering numbers into the DVS service. There is a 1 in 10 chance that the check digit will be correct.

Even if a person were to view client information by randomly entering a reference number with the correct check digit, the only identity information that should be displayed for any individual would be image, name and date of birth. No contact information (such as address, phone number or client's email address) would be available from the DVS.

The only impact that this proposal would have on clients is that they must use either a driver's licence, adult age card or an Australian identity card to identify themselves in order to receive government services. A shift in governmental policy would involve considering that the client has given implied approval to check their proof of identity document (and subsequently the client's identity) that they have provided in order to prove their identity.

For the majority of people in the community, using a driver's licence as proof of identity is current practice and they are unlikely to change. According to the proposals presented in this book, clients can choose which product they wish to use to prove their identity. Nevertheless, it is unlikely that many clients will choose to obtain the Australian identity card because there is no additional functionality or benefit from the card when compared to a driver's licence. Therefore, for the majority of people there will be no change to the current process of managing their privacy.

As mentioned above, on the whole, the driver's licence product and the adult age card display the same information, have the same card securities and use the same card production equipment and processes. The only real difference is that the adult age card is not used for driving. The same personal information, product securities, equipment, policies and processes for a driver's licence should be used to produce and manage the Australian identity card. This means client privacy issues would be the same for this product as for drivers' licences.

All state and territory transport authorities should be encouraged to update their driver licence and adult age card to be smart chip products so that they contain the same encrypted data as the Australian identity card and Queensland driver's licence. Once this has occurred, it is proposed that checking a client identity through the national DVS should take place automatically whenever the smart card is placed into a smartcard reader. This change is intended to speed identity checking and further ensure that a DVS officer is actually 'in possession' of a card product and not using a copy of the identity product. This further restricts possible unauthorised access to client information by DVS operators.

National Image Database

This book covers a proposal to establish a national image database that contains images of every client who has been issued with a driver's licence, adult age card or Australian identity card. However, client information such as the person's name, address or card number should not be available to the image database.

As a result, if unauthorised access to a client's image in the database occurs there should be no way of knowing who the person is, how to contact them or which of the three products the image belongs to. This information should only reside within state and territory systems. This practice is to ensure that it is very difficult for a criminal to steal someone's identify because it would require them to break into both the national service and then the appropriate state or territory government system, which is unlikely.

EIGHTEEN

Australian Privacy Principles

In Australia the collection and management of personal information is controlled by and consistent with the federal *Privacy Act* of 1988. The *Privacy Act* ensures government agencies collect and protect the community's personal information in a manner that is consistent with the Act.

The Act defines personal information to include: a person's name, signature, address, phone number, date of birth, medical records and bank account details.

To ensure that a consistent approach to handling personal information is used, the Australian Privacy Commissioner provides 13 Australian Privacy Principles to guide organisations in the management of personal client information.

A strategy in the proposal to introduce an Australian identity card to the community is to mirror the processes and policies governing the management of personal identity when clients apply for a driver's licence or adult age card. These processes and policies have been in place for an extended period of time and are well-tested. They are consistent with the Australian privacy principles and, most importantly, are trusted by the community.

To establish the proposed Australia identity card product, legislation would be required to formalise the federal government as the owner of the Australian identity card. Establishing service level agreements between the federal government and states and territories may allow the new product to be quickly introduced using existing resources and infrastructure with minimal change to state or territory based legislation.

The 13 Australian Privacy Principles and a brief response demonstrating how the proposals in this book fit each of the principles are listed below. The Privacy Principles are provided to support compliance with the Privacy legislation.

APP 1 – Open and Transparent Management of Personal Data

As mentioned above, the same personal identity requirements should apply for a person who applies for an Australian identity card as are currently in place for an Australian driver's licence or an adult age card.

In the case of the Australian identity card, the client should be advised that they may use the product as proof of identity to receive government services or benefits. At that time the client should also be advised that if they use the product as proof of their identity, information may be provided to validate and verify the integrity of the product and their identity.

APP 2 – Anonymity and Pseudonymity

State and territory transport authorities issue driver licences and adult age cards in the name of a person's official identity. Clients should be advised that government organisations may only deal with them in their official identity which is the person's identity that is recorded with the Registrar of Births, Deaths and Marriages or the Department of Immigration and Border Protection.

In the case of the proposed Australian identity card, it is intended that it should only be issued in a person's official identity. In this way, individuals are required to provide the same range of documents they would use to prove their identity in order to be issued with a driver's licence or adult age card.

It is proposed that all three identity card products will contain a smart chip. In the case of low-risk business transactions, it may be possible for the client to have an identity reference for them recorded by inserting the smart chip into a card reader but without providing any of their actual identity information. This transaction would largely occur as an anonymous transaction, and hence would not be associated with their identity.

In this way, an identity reference is recorded but not the person's actual identity details. If an event occurs and the person's identity is needed, the reference could be used by an authorised officer, such as a police officer, to obtain the person's full identity. An example of this model is provided later in this book.

APP 3 – Collection of Solicited Personal Information

State and territory transport authorities require a range of original source documents to establish a client's identity in order to issue them with a driver's licence or adult age card. The focus of these documents (and the transport authority process) is to prove:

1. the existence of an official identity
2. the existence or use of the identity in the community
3. that the person presenting is the owner of the identity.

To establish an individual's identity, transport authorities collect the person's name, date of birth, gender, address, person's height, eye colour, hair colour, facial image and signature. The relevant documents the person has provided to the transport authority to confirm this information are recorded and, in some cases, stored as scans.

The processes to obtain the proposed Australian identity card are designed to have the same requirements as the other two core identity products.

APP 4 – Dealing with Unsolicited Personal Information

Identity information collected by a state or territory transport authority about a particular individual is provided by that person. On occasions, transport authorities also receive information about a person that has been provided by police. This may include information that an individual has obtained official documents in a fake or alias name.

APP 5 – Notification of the Collection of Personal Information

State and territory transport authorities only receive personal identity information from clients when they present in-person. In cases where source documents that are being used to establish a person's identity are considered to be suspicious, transport agencies may undertake further checks with the relevant document-issuing agency or refer the matter to police for their investigation and advice.

APP 6 – Use or Disclosure of Personal Information

Legislation is in place to control the use and disclosure of personal information by state and territory transport authorities. Amending the state and territory legislation to include the Australian identity card would mean that the same provisions used for drivers' licences could be used for the management of personal information that is collected for the Australian identity card.

However, it is important to note this proposal suggests that transport authorities could provide driver licence, adult age card and Australian identity card information to the federal government's Document Validation Service.

APP 7 – Direct Marketing

Transport organisations undertake little or no direct marketing to clients that is not directly related to the products they have been issued to a client. Client information collected via the driver licence and adult age card is not provided to third parties who may be engaged in direct marketing.

APP 8 – Cross-Border Disclosure of Personal Information

Transport authorities have legislative approval to share client identity and driving information with each other as well as with police services across the country. Legislative changes would be needed to support the Australian identity card in this regard.

APP 9 – Adoption, Use or Disclosure of Government-related Identifiers

There is no proposal to remove any existing identity documents from the Australian market. There is also no proposal for a person to be known by only one identifier across multiple organisations (which is the current situation).

In cases where a person holds a driver's licence and adult age card, the same client reference number that is displayed on the document may be recorded. In the case of the Australian identity card, a different client reference number may be necessary. This matter should be considered by the federal government in consultation with states and territories.

It should be noted that in the case of drivers' licenses there is no national numbering system in place for either driver's licences or client references. Each state and territory generates their own client reference numbers and driver's license numbers independent of each other and other systems.

From a privacy perspective, it may be preferable to continue with this practice of issuing independent state and territory based client and product reference numbers, rather than use a national client numbering system such as the federal government's Medicare numbering system.

APP 10 – Quality of Personal Information

Drivers' licences held by individuals in the community are amongst the most up-to-date personal identity documents in use. This is because the driver licence product is supported by legislation that requires a client to inform the transport authority when they formally change their name or their address. In the case of Queensland, this process must legally occur within 14 days of the change.

It is proposed that the Australian identity card have the same legislative requirement for a client to update a change of name or address on the card by notifying their transport authority within a brief specified period.

The Australian identity card should be supported by the same document update policies as the driver's license product.

APP 11 – Security of Personal Information

Currently, transport authorities have protocols and systems in place that secure personal information which has been collected for driver licensing and adult age cards. It is proposed that the same standards, protocols and systems be used for the new Australian identity card.

APP 12 – Access to Personal Information

Driver licence and adult age card personal data is considered to be restricted data. No change is proposed to this policy and the Australian identity card should have the same policies applied to it.

APP 13 – Correction of Personal Information

Transport authorities maintain up-to-date and correct information on their clients. There is no change proposed to this policy and the Australian identity card should have the same policies applied to it.

NINETEEN

Objectives of the Australian Identity Card

Australia has a long history of attempting to introduce a national identity card. The first and most controversial attempt was made in 1985 under the Labor government by the then Prime Minister Mr Bob Hawke. Mr Hawke proposed introducing a national identity card called the Australia Card. At the time, the Labor government did not control the Senate and the proposal was defeated.

This was a significant proposal because it included a mandate that Australian citizens must obtain the Australia Card. The initiative was designed to reduce the incidence of tax avoidance and as a mechanism to reduce fraudulent transactions that were mainly related to government health and welfare payments.

Importantly, the Australia Card proposal also contained penalties that could be applied to businesses who failed to use the Australia Card, such as for identity purposes when employing new staff. The proposal also allowed the government to freeze the payment of Social Security benefits to clients who failed to show the Australia card document to concerned government agencies.

Accordingly, the Australia Card proposal generated a significant backlash in the community due to what many people may have considered to be unnecessarily harsh requirements and penalties.

Since the concept of an Australian national identity card was first proposed, a number of Australian governments (both Labor and Coalition) have indicated that they would support the introduction of a national identity card. The most recent support came from the Coalition government in 2006 when Mr Joe Hockey proposed to expand the existing Medicare card. It was proposed that this document would be called an Access Card but, once again, it contained many of the features of the initial Australia identity card proposal.

The proposals in this book are aimed at providing a national identity card to Australian citizens as an optional personal identity document. It is not proposed that it become a mandatory requirement to obtain the card. However, it is proposed that a person proves their identity to receive a government service or benefit. This could be done using either a driver's licence, adult age card or an Australian identity card as primary identity documents.

The proposed purpose for the Australian identity card is to provide an official government-issued identity document that is nationally recognised and can be validated. This product may appeal specifically to non-drivers such as the elderly. However, the government should also consider making the product available to citizens who are under driving age, such as teenagers. This would offer them the same level of identity protection as adults, as well as providing a nationally recognised identity document.

The Australian national identity card is likely to be more attractive to teenagers if the product is smart chipped and contains an electronic identity that supports online systems and interactions.

Australian citizens should, however, be able to continue to use identity cards that are issued by individual organisations to obtain services from those organisations. In this case, one of the three core identity products might to be used to initially identify or register a person as a client of an organisation. Once registered, an organisation might issue a person with the organisation's own identity card for the person to use to obtain services in the future.

There is no proposal for private sector organisations to comply with the requirement for clients to use a driver's licence, adult age card or Australian identity card as primary proof of identity. Nevertheless, private organisations may benefit from a policy change were initial proof of identity requires the production of one of the three core identity products. These products would be available for checking through services such as the DVS. This may be especially attractive for organisations who manage high risk transactions, such as banks.

Big Brother Concept

When the initial Australia Card proposal was first raised some segments of the Australian community (particularly privacy groups) likened the Australia identity card proposal to George Orwell's Big Brother concept that was presented in his well-known book called *1984*.

In *1984* Orwell talks about an entity called 'Big Brother' who supposedly looks after the citizens of a community. Big

Brother is not a single individual but, instead, the ruling party of a fictional country. Allegedly, Big Brother looks after the wellbeing of its citizens by constantly monitoring their movements, but in reality the monitoring was solely for the benefit of the ruling party to maintain control over the community.

Today, the term 'Big Brother' often refers to higher levels of community surveillance by government agencies. Potential abuses of official government power, including deceiving the community are linked with higher surveillance practices.

Scaremongering in relation to the Australia Card proposal and its possible link to the Big Brother concept by privacy groups in the mid-80s was effective. Many Australians considered the Australia Card to be an unnecessary interference by government and an intrusion into their daily life. However, it is unlikely that this same approach of attempting to scare the general community would be as successful these days.

At this point, it should be noted that identity management legislation may not have been fully in place in the 1980s which would have meant that the community may have been at risk from potential misuse of their information by government. These days there are formal institutions, standards and practices in place to better safeguard the Australian community's privacy and data.

The Australian community is now also more educated about the need for better identity security and the varying technologies that can enable this to occur. The community is also generally aware of the threats and costs incurred through identity crime and may be more supportive of governmental initiatives that aim to protect them from this burgeoning problem.

Impact on the Australian Community

Most of the proposals in this book are not controversial and may be considered systemic improvements that have no direct impact on the community. However the concept of an Australian identity card is an important policy decision for government. The identity card proposal is expected to generate media attention and public comment.

By closely linking the new identity card to the driver licence product and related processes, it may reduce community concerns about the Australian identity card proposal.

Clients are already familiar with the process and requirements that occur in a transport authority office to obtain a driver's licence. The intention of this proposal is to mirror identity components of the driver licence product, including engaging transport authorities as service providers. Having the Australian identity card as an optional product may also reduce many of the community concerns with a national identity card.

This proposal is designed to cause minimal impact on the majority of Australians who are drivers. Most Australians already use the driver's licence product (or, in the case of non-drivers, an adult age card) as their main proof of identity document. No change is expected or intended to this practice.

The proposal to ensure the Australia identity card is not compulsory means that driver licence holders can continue to use these products as their primary identity document with no change, no cost and no need to take any action. The non-driving adult community (which is typically the elderly) would have the option to obtain the new Australian identity card which may be more appropriate to their situation or continue to use the existing adult age card product as their primary identity product. In addition, the federal government may choose to extend identity protection to citizens who are under driving age.

Not requiring the Australian identity card to be compulsory is a key difference with this proposal to previous national identity card proposals (including the Hawke government's Australia Card proposal). The decision to obtain the identity card should be a personal choice.

In addition, there should be no requirement to show the Australian identity card even if a person has obtained the card. A person could continue to show their driver's licence for some transactions and choose to show their Australia identity card for others.

There are also no penalties proposed other than those that relate to failure to update the Australian identity card, such as when a person changes their name or address. These would be the same conditions as for the driver's license.

By not making it mandatory to obtain the new Australian identity card, many of the negative 'Big Brother' related reactions from the community and privacy groups that occurred in the 1980s are expected to be minimised. The Australia identity card proposal outlined in this book may be considered to be a 'non-event' for the majority of Australians and therefore more likely to be readily accepted.

PART FIVE

Enhancing the
Core Identity Products

TWENTY

Smartcard Driver Licences and Adult Age Cards

All state and territory transport authorities should be encouraged to transition their driver's licence and adult age card products to become smartcard products. In addition, space should be made available on the smart chip to allow governments and possibly private sector organisations to store data.

Some data such as client identity information should be placed on the smart chip by government as a mandatory requirement. In other cases, clients should be permitted to choose if they want data from other organisations to also be placed on the smart chip of their identity product.

It is suggested that the mandatory data placed on the smart chip include the following details:

- card reference number
- personal details such as full name, date of birth and gender

- an application that supports the transmission of encrypted data with the person's online identity such as currently occurs with Public Key Infrastructure (PKI).

The card reference number and the client's personal details should be encrypted on the identity product but offer different access levels. The card reference number should be automatically available when the card is inserted into a card reader with the appropriate minimum access level. This practise will ensure there is continued anonymity of some in-person transactions or activities. A person's details should also be available when the card is inserted into a smartcard reader that has a higher access level in the case of transactions in which the person's identity is required. It is possible for the application that manages data encryption to be physically activated by an individual using a Personal Identity Number (PIN).

In the scenario above where only the card reference number has been collected, the organisation which collected the card number would not be aware of an individual's identity. However, in situations in which the identity of a person was required, an authorised person (such as a police officer) would be able to identify the person using the card reference number.

Under this proposal, mandatory information could be added to the smart chip before it is issued to the person or cardholder. Each individual could then select additional applications or information they wanted added to their driver's license or other core identity product. Organisations, such as banks, who might want to add their organisation's information to a smart chip product should initially be approved by government prior to being allowed to access the smart chip to add their application and data to the person's identity product (and only if the person chooses to activate it).

Shifting the Australian community to a reliance on the three identity documents in the future will allow some manual transactions to be transitioned to online transactions and services. This, in turn, may strengthen monitoring and control of these transactions.

For instance, once the majority of Australians possess a smartcard identity product, a change to the DVS should be implemented. This is to ensure that the smart chip is exclusively used to access a client's identity information through the DVS without the need for manual data input by an operator. This is to reduce the opportunity for unauthorised access to a person's personal data by a DVS operator.

Below are a variety of examples that illustrate some of the benefits that may be possible by using a smart chipped identity product:

1. The federal government might use the smart chip on the three core identity products to replace the current Medicare card. This means that a Medicare application would reside on the smart chip of the core identity products as a smartcard application and a separate Medicare card would not be issued.

2. Use the smart chip to reduce access to poker machines by problem gamblers. In this example, the government or the client may request that access to poker machines that are present in licenced venues be reduced. By requiring the smart chip to be inserted into a card reader located within the poker machine area, a gambler's access to the area may be restricted if necessary.

3. Monitoring the purchase of some types of over-the-counter drugs. For example, using the smart chip as an integral part of the purchase transaction at a chemist

may quickly identify clients who are buying excessive amounts of particular drugs (such as codeine). This drug is sometimes used in the production of illegal drugs.

4. Linking the purchase of gun ammunition to the gun calibres registered to a gun owner. In this example, a gun owner who is registered with a .22 calibre gun would not be entitled to buy ammunition for a .243 calibre gun. This action might restrict the use of unregistered firearms within the community.

In the fourth example outlined above, the approach involves registering the gun owner's license and authorised firearms on the smart chip. As a result, the gun owner would only be authorised to buy ammunition that matches the calibres of his registered firearms. Furthermore, the two documents that are currently issued by government – (1) a Gun license and (2) a list of approved firearms – could be replaced with a government smartcard application that would reside on the identity products.

Currently, there are almost 1 million gun owners and approximately 3 million registered guns in Australia. But according to a 2016 report by the Australian Criminal Intelligence Commission, it is estimated that there are over 250,000 illegal and unregistered firearms in Australia. Restricting the purchase of ammunition that can be used with unregistered firearms may reduce their value and also the prevalence of unregistered firearms in the community.

Implementing this approach would result in increased safety levels, reduced costs and improved efficiency if it is no longer necessary to issue two separate documents (i.e. gun licence and record of registered firearms) to gun owners.

Above are some simple examples of how existing services and products could change with the widespread introduction of a new form of technology available to all adult Australians. In some cases (such as the gun license example described above) the government could mandate that specific product information would be held in the smartcard of an individual's identity product.

In other cases, the community would need to be convinced of the benefits of adding information to the smart chip of the identity product. This could involve an individual making the decision to replace a physical product, such as a credit card, with a smartcard application that resides on their core identity product.

For non-mandatory services, the marketing message to Australian citizens could be for them to choose whether or not to allow the placement of an organisation's application onto their smart card product. This marketing would occur with the knowledge that the government has already 'enabled' the community by providing the new technology platform to its citizens.

Enhancing the three core identity products with smartcard technology and providing an electronic identity to Australian citizens as a mandatory requirement offers public and private sector organisations the opportunity to provide new services and change existing services in a way that may not be possible without the new technology.

For instance, in this book I have only touched on the possible opportunity to use an electronic identity to improve the behaviour of some people who use social media services. While online interactions through social media should still occur anonymously, users could interact online with an awareness that their identity could be sourced by an authorised person (such as a police officer) if necessary.

Certain countries (including Germany) are concerned about the activities of some people who use social media services such as Facebook to distribute hate information, fake news and terrorist-related material. To manage this concern, the German Cabinet has approved the use of heavy financial penalties to social media companies who fail to quickly remove such inappropriate information from their online services. In response to Germany's decision, Facebook stated that it will employ over 700 people who will be tasked with reviewing and removing inappropriate material.

Rather than funding 700 staff to monitor social media submissions a more effective way to solve this issue is to discourage the posting of inappropriate material on social media in the first place. It is unlikely that users of social media services will act inappropriately (such as distributing inappropriate images and bullying) if their identity can be quickly traced by an authorised person.

If social media services were modified so that they required a formal government-issued identity to be used when accessing the service (such as providing the person's electronic identity from a government-issued smartcard) the problem may be solved.

While such enhancements may sound far-fetched, in Part Six of this book I will discuss the current rollout of smart card enabled identity cards across all European Union (EU) countries. These new identity cards contain the cardholder's electronic identity provided specifically for the purpose of being used for online interactions. Potentially, the 510 million EU citizens may be 'enabled' to use this type of enhanced social media service if it became available.

Obviously, these potential applications for improved secure smartcard interaction need further development and would require supporting infrastructure and systems.

Contact/Contactless Smart Chips

Smartcard chips are available in three formats:

1. contact
2. contactless
3. hybrid.

Contact smartcards require the card to be physically inserted into a smartcard reader in order for contact to occur between the reader and the smart chip. Once the reader makes contact with the smart chip, the stored information can be read or written to the smart chip. The Queensland driver's licence contains a contact smart chip.

Contactless smartcards require the smartcard to be placed in the proximity of a smartcard reader. The smart chip is embedded in the plastic card which also has an aerial implanted in it. The aerial is designed to manage transmissions between the smart chip and the reader without the need for physical contact between the card and the reader.

Hybrid smartcards contain both contact and contactless smart chips. The card provider can determine which applications or transactions can occur in the case of each type of smartcard.

Credit cards issued by Australian banks are usually hybrid smartcards and contain both contact and contactless smart chips. This gives clients the option to insert the credit card into a smartcard reader and enter a Personal Identity Number (PIN) to process a transaction or, alternatively, to place the card in the vicinity of the smartcard reader. In this case, using the contactless chip means no PIN or signature is needed to make a purchase. Not surprisingly, there have been reports in the media of spikes in credit card fraud as a result of criminals using stolen credit cards and the contactless chip technology to make unauthorised purchases.

In the case of hybrid credit cards, financial institutions have the option to set a parameter whereby the contactless smart chip can be used for low value transactions. However, the contact smart chip (which must be used with a PIN) is used only for high-value transactions.

In terms of the three core identity products, it is proposed that the government should consider contact smart chips only which require the card is to be inserted into a smartcard reader in order for contact to be made with the smart chip. This may reduce community concern that an individual's data could be accessed from the smart chip without them being aware of it.

Potentially, unauthorised access to a person's data that may be held on a contactless smart chip could occur in situations such as in a crowded street or crowded train carriage. This occurs when a portable smartcard reader is moved close to a person's wallet or handbag in order to extract data from the chip without the person knowing.

The potential for this crime has already been highlighted in 2014 by Australian Police in relation to criminals 'electronically pickpocketing' credit card details from a person's hybrid credit card. Criminals are then able to use credit card details to make online purchases. This is possible because some banks did not encrypt the bank credit card number or expiry date on clients' cards.

TWENTY-ONE

Liability Concerns Associated with the Australian Identity Card

An issue that is rarely discussed that is important to consider in the early stages of introducing an Australian identity card to the community is the potential for liability concerns arising from the incorrect issue of the card. Currently, the driver's licence product is the community's main identity document; however, this document is issued by transport authorities for the purpose of driving and not for personal identification in non-driving situations.

The fact that the majority of people in the community choose to use the driver's licence as proof of their identity in most situations is entirely their choice. Transport authorities are able to distance themselves from any legal responsibility involved in incorrectly issuing a driver's licence to a person who then subsequently uses it for fraudulent purposes. Their rationale is that the driver's licence is issued for the purpose of road safety and not as proof of identity in non-driving situations. Similarly, an adult age card is issued for the purpose of proving a person is over 18 years of age and legally able to enter pubs and clubs that serve alcohol and not for other proof of identity purposes.

Therefore, litigation to recover monies lost where a criminal has used a fraudulently obtained driver's licence or adult age card as proof of their identity are unlikely to succeed.

The community is increasingly reliant on stronger identity management that primarily involves the driver's license and adult age card products, and systems such as the Document Validation Service that validate the driver's license product. This has meant the ability for state and territory governments to continue to distance themselves from responsibility in the identity management business is becoming less certain.

Providing an Australian identity card should include consideration of legal responsibility for incorrectly issuing an identity product to an individual who then uses it to fraudulently misrepresent themselves. Legislation may be necessary to protect the government from claims made in cases where the Australian identity card has been fraudulently used.

Alternatively, the federal government could adopt the same approach as state and territory transport authorities by assuming the position that the Australian identity card is issued only for the purpose of proving identity to obtain government services.

Private organisations can then choose whether or not to accept the Australian identity card as proof of identity by clients who seek to obtain their services without receiving any government assurances or responsibilities concerning the client's identity.

Queensland Government & Identity Scanning

In previous chapters in this book, I discussed the strong penetration of the driver's licence in the Australian community and its use as the community's main personal identity document. As mentioned above, while the driver's licence is issued solely for the purpose of driving a vehicle, the government is increasingly expanding its involvement in identity management.

An interesting example of the growing prominence and expansion of identity management in the Australian community is the Queensland government's 2016 initiative to reduce alcohol-fuelled violence.

To reduce late-night violence by intoxicated people, the Queensland government introduced new laws and also conducted a trial that restricted the serving of high alcohol drinks after midnight and lockout laws at licensed venues that commenced at 1.00 am. The lockout law meant patrons could not enter a licensed venue after 1.00 am.

These laws turned out to be unpopular with pubs and clubs, particularly in entertainment precincts. Industry representatives believed they were losing money as a result of the laws and indicated that they would campaign against these laws in an upcoming Queensland government election.

As an alternative, the industry proposed the mandatory scanning of identity documents (which would primarily be drivers' licences) of potentially everyone who enters a pub or club. Using information obtained from clients' scanned identity documents, those patrons with poor behaviour could then be banned from pubs/clubs.

At the time of writing this book, the proposal to scan patron's identities was being considered by the Queensland government. If the government agrees to the industry proposal, it would be a further shift towards government officially recognising the driver's licence as an identity document in non-driving situations. The significance of this proposal is far-reaching in that, in theory, people would need to identify themselves every time they enter a pub or club and their core identity product would be scanned.

A valid concern linked to the industry proposal of scanning driver's licences and other identity products at pubs and clubs is that these organisations may potentially create large databases of client personal details in some cases, including home addresses.

Access to these scans would make it easier to create fake documents using the details of real people which, in turn, may actually increase the levels of identity crime and identity theft already occurring in the community.

If the personal data of patrons was to be held locally at a pub or club, it would take a significant 'leap of faith' to believe the local pub/club will prioritise and provide long-term funding to ensure high levels of data security of this personal information.

It may prove to be considerably more cost-effective for the pub/club to apologise after a data breach occurs than to fund data security requirements on a long-term basis. Similarly, it will be easy for a government to chastise the pub/club after the data breach occurred for not complying with data security obligations. However, the unfortunate outcome would be the loss of patron's personal data that may then become available for sale to criminals. Ultimately, patrons of these pubs/clubs may incur fraud-related financial loss as a result of a data breach.

Alcohol-fuelled violence is a national problem. There is a potential for the proposal put forward by industry to be adopted in other states and territories. If pubs and clubs are able to store full customer details (including the facial image and a customer's address) it creates an opportunity to:

1. Stalk people (particularly women) in their home

2. Make it easier for retribution to occur at a person's home as a result of an incident that took place at a pub/club

3. Provides an opportunity for pub/club staff to be aware that cardholder homes may be vacant which, in turn, may mean that they are more likely to be burgled

4. Gives the pub/club (or another third party) the ability to directly market their services to cardholders at their homes.

It is important to note that, in the case of the Queensland driver's license, a person's address is recorded on the back of the driver's license. Consequently, the person's address will not be included in a scan if only the front side of the licence is scanned. However, other Australian state and territory drivers' licenses display the address on the front side of drivers' licenses.

Most people who attend Queensland pubs/clubs hold a Queensland driver's licence that contains a smart chip. The smart chip contains a card identity number.

An alternative proposal to pubs/clubs scanning Queensland driver licences and creating databases of Queenslanders' personal information is for the pub/club to instead use a smartcard reader to record the smart chip reference number. This means that pubs and clubs would not have access to patron's personal information.

The pub/club would store the reference number with the date and time the smartcard was read. If the pub/club wishes to have a person banned, they could check their own CCTV records to see the time the person entered their pub/club and then match this information with the time the person used their smartcard to enter their premises.

The pub/club should then be provided with the cardholder's image online from the Queensland Department of Transport and Main Roads, who have the cardholder's facial image from their driver's license or adult age card. This will allow staff at the pub/club to confirm the person they wish to ban from licensed venues.

Under this model, the process of banning a person from entering a licensed venue can occur without staff ever recording the person's name or other identity information. There is also no need for pubs/clubs to purchase and manage data security systems and protocols. This model maintains client privacy while minimising data security risks.

When a driver's licence or other core identity product is inserted into a pub/club smartcard reader it should do an automatic online check using the person's smartcard reference number to confirm the person has not been banned from entering the premises. This automated process should occur faster than the time it takes to scan a document and, therefore, should have less impact on the operations of pubs/clubs.

However, it should be noted that, under the industry proposal, an ID scanning system for driver licences may still be required for non-Queenslanders. Interstate and overseas visitors do not have smartcard driver licences and may need to provide their full details through a driver licence scanning process which may contain the risks described above.

TWENTY-TWO

Australian Identity Card and Fingerprinting

Over time, there has been some comment by a number of politicians about using fingerprints to identify Australian citizens who seek government benefits. Once again, this suggestion is designed to reduce fraud by improving identity management when people apply for government benefits. Recently, an Australian Federal politician has proposed to link citizens' fingerprints to a possible Australian identity card.

Smart chip technology is well-suited to support the use of fingerprint scanning in the delivery of high-risk transactions. A scan of a person's fingerprint is taken and an algorithm is used to map the fingerprint features. The algorithm is then compared with the fingerprint algorithm that is stored in the smart chip which either confirms or rejects a match with the person.

Although fingerprint scanning technology and the use of smartcards to support such a service utilises mature technology, the process of fingerprinting of individuals is quite intrusive for the community. It also involves a large service delivery cost as clients would need to attend government offices to have their initial fingerprint (or fingerprints) registered on their smart chip identity product.

In Australia, fingerprinting is associated with criminals so, consequently, the general community is expected to strongly resist being fingerprinted. It is very likely that an attempt to fingerprint the Australian population as a whole (or certain members of the community such as the recipients of government benefits) would generate a significant backlash by all segments of the community.

Since the 1980s, rather than fingerprinting people, state and territory transport authorities have instead printed facial images on driver licenses to counteract identity crime. Today's facial recognition technology is an unobtrusive biometric technology with the added advantage of having little impact on service delivery speeds.

Providing the client's image online from a secure source (such as transport authorities' databases to the new DVS) may achieve the same outcome of secure identity verification without the need to fingerprint people. Using the existing facial biometric (image) that is held by state and territory transport authorities avoids the need for clients to take any action. This development using the existing image could occur immediately as an enhancement to existing government systems with no impact on Australian citizens.

PART SIX

Overseas Systems for Identity Management

TWENTY-THREE

Identity Management in Australia's close neighbours: New Zealand and Singapore

New Zealand

New Zealand has enjoyed a special relationship with Australia for many years. This relationship includes exemptions from standard residency, citizenship and visa requirements to visit (and sometimes stay in) Australia that are not currently available to people from other countries. The only requirement for travel between the two countries is that an individual holds a valid passport. Consequently, there is a high level of migration and tourism between both countries.

Australians and New Zealanders who migrate between these two countries typically use their passport and their driver's licence as their primary identity documents, as well as for the purpose of establishing their driving credentials.

Like Australians, the New Zealand community use their driver's licence as their main personal identity document.

New Zealand's identity management processes for driver licensing are almost identical to those used by Australian states and territories. New Zealand provides representatives to the Austroads subgroup that has been formed through a collaboration of Australian transport authorities. Many of Australia's 'best practice' standards that have been developed through the Austroads organisation were also provided to the New Zealand transport authority for potential use.

It is suggested that if the proposals in this book are implemented then New Zealand should be offered the opportunity to interact with Australia's NEVDIS (and possibly DVS) and the proposed facial image database. This may strengthen Australia's and New Zealand's identity management practices in relation to Australian and New Zealand drivers visiting each other's country, thereby reducing the potential for fake driver licences.

Singapore

One of the earliest nations to introduce a national identity card was Singapore via the *National Registration Act* that was passed in 1965. The Act requires all Singaporeans who are over the age of 15 years of age to obtain a national identity card as a compulsory requirement.

Singapore's national identity card is managed by the Ministry of Home Affairs. The identity card contains the person's name, race, date of birth, gender, country of birth, facial photo, client reference number, barcode, fingerprint, date of card issue and home address. The client reference number on the card is a lifetime number that first appears on an individual's birth certificate.

The national identity card is extensively used by public and private sector organisations within Singapore. The client reference number that appears on the card is also extensively used by government and private sector organisations as a community-wide reference for a client.

The Singapore government manages an online service called Singpass in which Singaporeans use their national ID number and a password to access over 200 government services that are available online. There have, however, been instances of fraud where criminals have fraudulently accessed benefits through the Singpass service.

In 2014 the Singaporean government started investigating a new E-identity service to replace the Singpass user ID and password process. The motivation for this change was concern about the rising incidence of identity crime (including identity theft) that was occurring in the Singaporean community.

In 2016 it was reported that the Singapore government was looking to trial a new electronic identity that would reside on a smart phone. A person's identity would be stored in a secure location on the mobile phone SIM card using smartcard technology. The person would then use a PIN to activate the electronic identity in the SIM card allowing encrypted messaging to occur that would support online government service provision.

An important factor in the Singapore example is the fact that a very high percentage of the Singaporean community own a mobile phone. Such a high level of mobile phone usage within the community has not occurred in many other countries. Consequently, this would reduce the effectiveness of replicating this initiative elsewhere.

TWENTY-FOUR

Identity Management in the United Kingdom and the European Union

United Kingdom

The United Kingdom (UK) serves as a great example of the excitement caused in the community around the issue of identity management and the concept of a national identity card .

The UK national identity card proposal arose from the need to better protect the community from identity crime. This need was recognised during the 1990s by the UK Labour and Conservatives parties. At this time both political parties publicly supported the introduction of a national identity card.

In 2006 the *Identity Card Bill* was passed into UK law. This provided the framework for the introduction of a national identity card that was linked to a national identity register and the collection of client biometric details. The biometric component of the UK proposal involved collecting and storing facial images, iris scans and fingerprints from the UK community.

Initially, obtaining the national identity card was intended to be voluntary by UK citizens unless a person applied for a passport. However, the government had decreed that obtaining the national identity card would become compulsory for all UK citizens from 2010.

The UK community was generally not supportive of the proposal. The main concern revolved around the creation of the national identity register which would be a single large database containing all the community's identity details. Having all the community's identity data in one system was seen as risky and open to privacy intrusions. Interestingly, there was much less community concern associated with a national identity card.

While the concept was easy to understand, the initiative struggled particularly in terms of funding. Changes were made to the initial model mainly due to funding limitations. For instance, a change was proposed whereby data would be stored within an existing Social Security database instead of establishing a new purpose-built database. Issues with data security, accountability, funding and governance were never resolved by the UK government which, in turn, would have diminished public confidence in the new national identity service.

However, in 2008, the first identity cards became available to the community under a Labour government. Demand for the identity cards by average UK citizens was not high. But, in 2010, the Conservatives won government in the UK and the legislation that supported the identity card scheme (including the national identity register) was repealed. The scheme was discontinued and the associated client data in the national identity register was destroyed.

Today, the UK community use their driver's licence (which includes the licence holder's facial image) and passports as their main physical proof of identity documents.

The UK government has, however, continued to invest in personal identity management which is largely focused on online verification of a person's identity. In May 2016 the UK government launched its GOV.UK Verify scheme. At the time of launch this scheme had engaged nine private sector identity providers. Their task is to identify UK citizens and to facilitate the provision of online access to government services. It is understood that the service providers take about 15 minutes to verify a person's identity and issue them with a user ID and password that allows them to access government services through the service provider.

This project was initiated in 2011 by the Cabinet office of the UK Government. It was planned to have up to 30 UK government departments providing online services to clients through the new service by the end of 2016; however, this target was not achieved.

Citizens in the UK can also still access services through traditional channels such as service centres and telephone if they wish. At this stage it is not known how successful the GOV.UK Verify scheme actually is. Moreover, the scheme is receiving some criticism due to the government's use of private identity providers with critics claiming that the government is attempting to outsource risk and blame for the potential mismanagement of the community's personal identity.

From the outset, it could be said that successive UK governments have mismanaged the issue of identity management. A consequence of failing to recognise and address at an early stage community concerns about data security and privacy has meant UK governments have created a higher degree of community resistance and mistrust about government-led changes to identity management.

There seems to be similarities between the UK and Australian governments (in relation to the Australia card initiative), in their failure to successfully manage the identity issue.

The European Union

Unlike the United Kingdom, the majority of European countries that make up the European Union have agreed to a proposal for a national identity card to be available to all citizens and have not wavered from that position. However although all European Union countries eventually were in agreement, initially not all countries actually offered their citizens the agreed European Union (EU) identity card.

In 2006 the EU agreed to a common design and minimum security standards for national identity cards. The cards were to contain the person's name, date of birth, gender, nationality, facial photo, signature, card number and card validity period. The EU identity cards were also designed to be machine-readable and so would incorporate either a magnetic stripe or smartcard technology.

Although EU nations agreed to the identity card requirements, the actual rollout of the card has not been consistent across the EU. Several EU countries such as Italy have only recently changed from a paper identity card which was a relatively low security document and susceptible to fraudulent reproduction to higher security card documents. Some countries including Austria, Belgium, Estonia, Finland, Germany, Italy, Liechtenstein, Lithuania, Portugal and Spain issue individuals an EU identity card which contains smartcard technology. The smart chip incorporates an electronic identity which the cardholder can use to authenticate themselves by means of a PIN.

Today, most of the EU identity cards contain smartcard technology and some also contain a bar code. No EU identity cards contain magnetic stripe technology.

The EU national identity card can also be used as a travel document in the same manner as a passport. The identity card is accepted as a travel document in EU countries as well as a number of nearby countries, including Egypt, Turkey and Tunisia. The national identity card is supported by a central database that validates the identity product and confirms a person is not under investigation for crimes by a EU country.

Initially, most European Union countries offered the national identity card as an option for citizens. If someone chooses not to obtain a national identity card, they can use their country's driver's licence as their primary proof of identity document.

Malta is a good example of a country that supports the EU national identity card. Maltese legislation mandates every person over the age of 14 years is to have a national ID card as a compulsory requirement. Unless an individual was under 18 years of age when they obtained the card, it is valid for a ten-year period. If they are under 16 years of age, they must update the card soon after their 16th birthday. The card then needs to be updated again soon after the person's 18th birthday.

The implementation of the EU's national identity card has been underway for over 10 years and is not finalised. An area in which the EU seems to be progressing more quickly is in the provision of electronic identities to citizens of the European Union; an initiative which is being managed through the national identity card process.

In July 2014 the EU Parliament passed a regulation supporting the issuing of electronic identities to all EU citizens as a mandatory requirement. All EU countries are mandated to commence the rollout of electronic identities to their citizens by September 2018. The electronic identity is to reside on the national ID card using smartcard technology.

This plan shifts the policy of obtaining a national identity card into being a mandatory requirement. It also ensures that the identity card will mandatorily become a smartcard product.

The focus of this initiative of providing an electronic identity to EU citizens will further support the operation of a single European market. While it does have security-related benefits, ultimately security is not the main objective of this initiative. Rather, the intention is to increase community trust to do business online within the EU and across national borders.

At this stage, the European Union is one of the world leaders with regards to providing electronic identities to community members. High levels of support for the provision of electronic identities would explain the transition by EU countries to smartcard-enabled identity cards

TWENTY-FIVE

Identity Management in United States of America (USA)

The United States of America has a very similar approach in its management of personal identity to Australia. The United States also runs a federated government system that contains a federal government and 50 independently governed states.

As is the case with Australia, the United States also has no national identity card. Historically, within the USA, attempts have been made to introduce a national identity card which have failed primarily because of concerns about privacy and the role of the federal government.

The main personal identity document currently used in the United States is the state-issued driver's licence. Like Australia, in order for an individual to obtain a US driver's licence proof of identity is required. This primarily resides with a birth certificate that is also managed and owned by state authorities. Like Australia, a birth certificate is not a primary proof of identity document; however, it is proof of citizenship and in the US it is called a 'feeder document'.

The birth certificate supports a person's application for a primary identity document such as a driver's licence or passport. The documents required by American transport authorities to prove identity to obtain a driver's licence closely match those required by Australian transport authorities.

Most US drivers' licences also contain client facial images which were introduced in the 1980s. The addition of facial images to drivers' licences was in order to support changes to liquor laws as a result of raising the legal age to consume alcohol to 21 years of age.

Incorporating a person's facial image on drivers' licences was designed to support identity checking in pubs and clubs that serve alcohol. The driver's licence is also used as proof of identity in electoral voting if it contains a client's facial image.

Typically, only the elderly or a person who objected on religious grounds could opt for a driver's licence without a facial image. This option of no facial image on the driver's licence was originally available in 13 of the 50 US states. However, with the stronger US focus on identity management during the past few years, it is no longer available as an option in any US state.

In comparison to Australia, the United States has a greater emphasis on national security and border protection. This has occurred primarily in response to the terrorist attack that occurred on 11 September 2001 in which terrorists hijacked a number of aircraft and used fake drivers' licences as identity documents to board the aircraft. Since that attack, the United States has significantly increased its focus on securing the identities of all Americans by improving American proof of identity documents and associated identity processes.

Like Australian states, US states also provide an adult age card. In the USA it is called a non-driver ID card. This identity card has the same acceptance within the community as a form of personal identity as a driver's licence. Both products are considered primary identity documents and are widely accepted by government and private sector organisations.

In 2005 the federal government passed the *Real ID Act* which was a direct outcome of the September 2001 terrorist attack. This federal government Act set the standard for the design and content of state-issued driving licences and non-driver ID cards.

The standard set by this Act closely matches the requirements of Australian drivers' licences in that it requires the mandatory addition of a client's facial image to all US driver licences. Furthermore, according to the Act, only machine-readable driver's licence and non-driver ID cards may be issued. This means either magnetic stripe or smartcard technology must be added to the driver's licence and non-driver ID card products.

All US transport authorities have transitioned to an Australian-style digital driver's licence with approximately 40 transport authorities using facial recognition technology within their image databases. In some jurisdictions, client fingerprints are also captured as part of the driver licence and non-driver ID cards processes.

US transport authorities coordinate their activities through the American Association of Motor Vehicle Administrators (AAMVA) which is the equivalent to Australia's Austroads organisation. Like Austroads, AAMVA has the same focus on national consistency between transport authorities within the United States and also with Canadian transport authorities.

The *Real ID Act* provides the US Department of Homeland Security with the authority to regulate state transport authorities in relation to their drivers' licences. The legislation also requires that all state transport authorities have access to each jurisdiction's client identity and driving information, including client facial images. Furthermore, the Act requires a full copy of client identity and driver licence information is to be stored in a single national database that is managed by the Department of Homeland Security.

Although the *Real ID Act* was established in 2005, some US states have resisted complying with the Act on the grounds that individual states are responsible for funding the costs associated with upgrading the driver licence and non-drivers identity card products. It is understood that there is also some resistance to the involvement of the federal government in what is considered to be a state managed process and product.

The US Department of Homeland Security has given the states until the end of 2017 to fully comply with the requirements of the Act. Failure to comply means drivers' licences from those states who have not agreed to the conditions of the Act will not have their drivers' licence products accepted as proof of identity for federal government business (i.e. to receive federal government services, to access federal buildings or to board aircraft).

The other important identity product used in United States is the passport card. The passport card is similar to a driver's licence in appearance and is used for international travel to specified countries in the local US region, such as Mexico and the Caribbean. The passport card cannot be used for identification when boarding a plane and is only valid for road and sea travel. The normal United States passport is used for air travel or to visit other countries.

Some transport authorities with jurisdictions close to border areas have incorporated the passport card into their updated driver's licence.

The US government also provides its citizens with a Social Security number that is a lifetime number to be used for tax purposes. These days, many parents apply for their child's Social Security number as soon as the child is born. This practice then allows the parents to access government benefits on behalf of their child.

While a number of government agencies use the Social Security number as a client reference number, it is not used as proof of identity. In recent years, some private sector organisations had started using the Social Security number as their organisation's reference number for a client. However, at the request of the federal government, this practice has since been discontinued. The intention is that a US citizen should not be known by one reference number throughout the community.

This is the opposite approach to that used in Singapore where the government-issued client reference number is used extensively by both government and private sector organisations.

Recently, US transport authorities, via the AAMVA, have been investigating the possibility of providing the US community with an electronic identity through the driver's licence process by adding smartcard technology to these products. This initiative is designed to support improved provision of online government services to the community, particularly by the states. At this stage it is unknown what progress has been made by transport authorities in terms of providing the US community with electronic identities.

TWENTY-SIX

Identity Management in the United Arab Emirates

I n the country of the United Arab Emirates (UAE), citizens and residents are fingerprinted as part of the process of obtaining a national UAE identity card.

UAE National Identity Card

The UAE identity card looks like the Queensland driver's licence whereby the person's digital facial image and personal details are displayed on the card which also contains contact and contactless smartcard technology. It is compulsory for UAE citizens to obtain a UAE identity card which is then renewable every two to three years depending on a person's residency or visa rights.

Residents of UAE attend a government office to apply for the UAE identity card. When applying for the card, their passport details are scanned as well as their residency visa which is contained in their passport. They are also finger-printed which includes all fingers, thumb and palm prints. Finally, a digital facial image is recorded and they pay a fee.

The UAE identity card application is processed as a back-office service. The client's facial image is stored in an image database and facial recognition technology is used to compare the client's current and previous facial images to ensure there is a match. A new image is also compared with other facial images in the database to ensure the person does not have a second identity.

Anecdotal evidence suggests that, initially, the community resisted obtaining the card as it was seen by some as being just an additional government tax. This mainly stemmed from the fact that, while it was compulsory to obtain the UAE identity card, it was not used for any purpose.

However, after some years, a policy change occurred whereby the UAE identity card was required to obtain most government services at federal, emirate (or state) and local government level. Many private sector organisations such as banks also adopted the requirement to use the UAE identity card before services such as opening a bank account would be provided.

The policy of linking the need to provide the UAE identity card with the provision of government services has dramatically reduced community resistance to the card. Consequently, there is now 100% compliance to the requirement to obtain and use the UAE identity card. There has also been some general discussion about adding UAE driver licences to the UAE identity card; however, this has not progressed any further.

Primarily, the UAE identity card is used as proof of identity to receive government services. However, from an operational perspective, client fingerprints are not used in any service delivery capacity except by the police when investigating crimes.

The UAE identity card smart chip contains client residency visa information and employer information. The three tiers of UAE government can read data from the UAE identity card when the card is inserted into a smartcard reader.

United Arab Emirates law decrees that government services can only be provided to a person with a current UAE Residency Visa. In the case of transport authorities (and probably other government authorities), inserting the UAE identity card into a smartcard reader allows client identity information to automatically populate the authority's database.

Government organisations can read information from the UAE identity card smart chip but cannot write information to the chip. Increasingly, private sector organisations require the UAE identity card before transacting with a client; however, they have no ability to access information on the smart chip.

Even though the UAE identity card is a solid personal identity product that has full market penetration in the UAE community it is still not always carried by community members. The card is carried specifically when it will be needed, such as opening a bank account or for first-time identity to new organisations.

Within the United Arab Emirates the most carried personal identity product is the driver's licence which is still used as proof of identity in most situations. The UAE driver's licence is a nationally consistent product and contains the same common personal and card information that appears on an Australian or a new US driver's licence. UAE driver licences also contain a digital facial image of the client which is stored in an independent emirate (the same as a state) transport authority database.

In the future, it is anticipated that the client's facial image will be provided to the transport authority for production of a driver's licence from the UAE Identity Authority. This Authority currently owns the UAE Identity Card. At that time, UAE transport authorities may discontinue photographing clients for drivers' licences and instead formally transfer full responsibility for identity management to the UAE Identity Authority.

Document Validation Service – Dubai Roads and Transport Authority

The country of the United Arab Emirates is made up of seven states that are called emirates. The most densely populated emirate is the Emirate of Dubai. The Dubai Roads and Transport Authority (RTA) is the organisation responsible for providing drivers' licenses and vehicle registration services within this emirate and is equivalent to an Australian transport authority.

The RTA issues a wide range of paper documents that relate to drivers' licenses and vehicle services – most of which are available online. This allows the RTA to offer 24-hour service delivery to clients from any location.

Common problems associated with paper documents (and particularly those that have been obtained online) are document validity issues and document quality issues. Paper documents are also highly susceptible to fraudulent production and alteration.

To address this problem, in 2010 the RTA established an online Document Validation Service (DVS) that is similar to the one established in Australia in 2009.

However, an important difference is that the Dubai Roads and Transport Authority's DVS only permits documents issued by the organisation's Licensing Agency that has primary responsibility for driver licensing and vehicle services, to be validated.

Any person can log onto the online DVS site and enter a document reference number that is shown on the paper document that has been issued by RTA. The DVS then displays a summary of the document and confirms that the information on the document is still valid.

If an individual requires more information they can access a general copy of the document. The specifics of the product that the document refers to will be displayed but the identity information of the person who owns the document will only be partially displayed. This reduces the ability for someone to make a fraudulent full copy of the real document.

Importantly, the image of the person that the certificate relates to is printed on the paper document and is displayed on the DVS screen when the online check occurs. This process makes it easy to cross check the person's image that appears on their paper document and the image that is displayed directly from RTA.

Therefore, while it is relatively easy to produce a fraudulent paper document or change a paper document, an online check with the document issuing organisation that shows the person's facial image ensure this process is relatively fail-safe. This, in turn, reduces the ability for identity crime to occur where paper documents are issued. The person's facial image is sourced from a copy of the facial image that was printed on the person's Dubai driver's license.

Dubai's DVS (which uses facial images to support identity management) highlights a major weakness of the Australian DVS.

It is important to note that, for cultural reasons, facial images of Dubai women are not printed on paper documents or available for verification through the RTA's Document Validation Service.

UAE – E-Gate Service

The UAE federal government also provides an E-gate service that manages passport control. This is a further excellent example of how biometrics can support improved business efficiencies even in managing high risk services.

Dubai is the most populated emirate of the UAE. Approximately 20 per cent of Dubai's population is made up of UAE nationals while the other 80 per cent are expats. Generally speaking, the Dubai community is wealthy and are frequent flyers. Dubai Airport (which is the home of Emirates airline) is the busiest passenger airport in the world. In 2016 Dubai Airport managed over 83 million travellers.

In 2002 the UAE government introduced the E-gate service which is a biometric-supported service that manages the passport process when UAE citizens leave or enter the country.

For a small fee, a UAE resident can choose to obtain an E-gate card, which is a smartcard that also shows the person's digital facial image and personal identity details. It looks similar to a Queensland driver's licence. For a slightly larger fee, a UAE resident can choose to have the E-gate card function loaded onto their UAE national ID card as a smartcard application, so they do not require a separate E-gate card.

Anybody entering or leaving the UAE undergoes a passport/immigration control process. Immigration/border control officers at Dubai Airport use the standard process of checking a person's passport to allow them to enter or leave the UAE that is used in most modern Western-style airports. However, due to the large volume of passengers and the manual checking process in operation at the airport, there are usually long queues of people waiting to have their passport checked to enter or leave Dubai.

Dubai residents who have an E-gate card can choose to enter or leave Dubai via the E-gate process. To leave Dubai, an E-gate card holder would follow the E-gate signs in the airport to a bank of electronic terminals that look similar to ATMs. They place their E-gate card on a terminal and their plane boarding pass under a scanner attached to the electronic terminal. The terminal reads the data from the E-gate smart-card and the plane boarding pass.

A back-end computer system determines if an individual is approved to leave the country, in which case a door opens and the person walks into an area approximately 1 metre x 2 metres. At this point, a computer requests the person to place a specific finger onto the screen which is then scanned. If the person's fingerprint stored on the smart chip of the E-gate card matches the person's identity, a second door opens meaning the person has completed the passport control stage.

This whole process typically takes about 30 seconds and means that the E-gate card holder does not have to deal with a queue of people. Likewise, it does not involve any assessment by an immigration/border control officer.

Entering the UAE through the Dubai airport is the same process for an E-gate cardholder – but without the need to scan an aircraft boarding pass. It again takes about 30 seconds and does not involve any assessment by an immigration/border control officer.

The E-gate service is extensively used by Dubai and UAE residents. The UAE's E-gate card can also be used by UAE nationals to enter or leave any of the six countries of the Gulf Cooperative Council (GCC); a process which occurs in the same way as described above and without the involvement of immigration/border control officers. The GCC is made up of the following countries – United Arab Emirates, Saudi Arabia, Oman, Bahrain, Kuwait and Qartar.

The E-gate service at Dubai Airport relies on biometrics and works brilliantly ... now if only they could speed up the baggage collection process!

PART SEVEN

Future Trends in
Identity Management

TWENTY-SEVEN

Australia's Capacity to Improve its Identity Management Practices

The processes associated with securing personal identity have become a priority for governments across the Western world. This trend is in response to the growing incidence of identity crime (including identity theft) that can be devastating to victims and exposes the entire community to rising fraud-related costs.

As a direct outcome of the Martin Place siege that occurred in December 2014, Australian governments are now well-positioned to strengthen the management of personal identity in the community through the development of new identity-related services that are linked to the National Identity Security Strategy (NISS). Historically, state and territory transport authorities have played an active role in securing the community's identity through improvements to driver licences and supporting systems, and they continue to provide this protection role.

Attempts to improve personal identity in many parts of the world (including Australia) have previously failed due to concerns about governments' abilities to secure personal data and maintain community privacy. However, the world is changing rapidly and hence securing the community's personal identity is now acknowledged by the Australian federal, state and territory governments to be a priority.

The task of securing personal data and privacy is still a major challenge for governments to effectively manage. These days the Australian community is more educated about the importance of privacy management and the costs associated with failing to protect personal identity.

Stringent privacy and data security provisions exist within Australian legislation that is matched with institutions that are responsible for overseeing identity management. Actioning the proposals in this book would better position governments to take a more active role in community protection through better identity management.

A low-impact approach to improving identity management by enhancing the integrity of existing identity products and continuing to utilise the driver's licence as a main identity product is proposed. Developments underway by the Australian Federal Government both support and strengthen the proposal that state and territory transport authorities continue to be the key providers of identity products to the community. In addition, it positions transport authorities to become the designated service providers of a national identity card.

At this point, there appears to be little evidence that the Australian government is committed to the provision of secure electronic identities to Australian citizens, which is a definite failing. Some Western governments are actively pursuing the provision of electronic identities to their citizens as a method for shifting high-risk, high-profile and high-demand government services to online delivery.

Online service delivery is intended to increase access to government services while, at the same time, reducing service delivery costs. Failing to provide the community with secure electronic identities that support online service delivery means there is a heightened risk of increased identity crime fraud (including identity theft) that is related to online services. This may cause mistrust about online security within the community and result in some citizens resisting the adoption of online service delivery systems.

A traditional problem for governments in providing online services to clients has been verifying the identity of the person who wishes to obtain the service. Providing the community with a secure electronic identity via the three proposed primary personal identity smartcard products is intended to solve this problem and support a shift to online transacting by government agencies.

Most governments that provide secure electronic identities to their citizens do so via smartcard technologies. Smartcard technology is currently the standard technology in terms of cost effectiveness, security and flexibility and is expected to be the primary technology over the next 20 or more years.

At this stage, it is proposed that the volume of primary smartcard-enabled identity products within the community be increased. For this to occur, transport authorities should be encouraged to convert all driver licence and adult age cards to smartcard products. The advantage of this form of technology is that it allows additional data to be stored in the smart chip while also increasing the integrity of identity products, thus making them very difficult to fraudulently reproduce.

However, an ongoing problem associated with the use of smartcards is the necessity for smartcard readers. This is particularly difficult if governments want clients to be able to access government services from their homes. To date, no government has adequately addressed this issue and been able to offer a community-wide solution.

Earlier in this book I recommended that government should not consider providing the community with smartcard readers but, instead, let individuals choose to fund the purchase of smartcard readers if they think there is value in this purchase.

With the rapid growth of smartcard usage worldwide, the cost of smartcard readers is currently already low and could be expected to fall further over time. Within the community, the cost of purchasing a smartcard reader is not considered to be a barrier in using smartcard technology with electronic identities.

As previously mentioned, countries in the European Union are currently rolling out smartcard-enabled national identity cards to their 500+ million citizens. Transport authorities in the United States are considering the provision of electronic identities to their 240+ million adult citizens which is expected to be supported by smartcard technology. Today most Australian credit cards contain smartcard technology and there is a smartcard reader in every shop that accepts credit card purchases. These developments are increasing the volume of smartcards and smartcard readers in use across the world.

It may be expected that through economies of scale and the increasing supply and demand for smartcards and smartcard readers that the cost of these technologies would decrease further over time.

As discussed previously in this book (in particular, the UAE's E-card for passport control and India's Aadhaar

Biometric ID Program), it is expected that governments will continue to invest in biometric technologies (including facial image recognition and fingerprint recognition) to better support improved identity management within the community. Biometric technology is well suited to being used in combination with smartcard identity products that Australia governments may be in a good position to introduce to the community in the future.

However, the final step in identity management would involve discontinuing the issuing of physical identity products such as drivers' licenses and, instead, relying solely on biometric technologies. In light of today's political environment and existing telecommunication technologies and capacities, this approach may be considered 'a bridge too far' for most governments. It may still be more than 20 years away for such a step into full reliance on biometric technologies for use with standard high-volume community applications to occur.

TWENTY-EIGHT

Future Impact of Smartcard Technology on Business Outcomes

Future developments in smart chip technology may also support a growth in online transacting with business organisations as well as government agencies. In terms of identity management for businesses, the primary objective for smartcard technology is to provide the business community (including government agencies) with a higher degree of confidence that a person is who they say they are. This can be achieved by providing better quality and more secure personal identity documents, by ensuring that the document can be easily validated and the person presenting the document can be verified.

Another desirable outcome for business organisations is to ensure the above objectives are implemented quickly and cost effectively. This can be achieved by using the community's existing identity practices and leveraging additional benefit from previous government investments in identity management.

Compressing the current range of primary personal identity products used within the Australian community to the following documents: (1) Driver's licence (2) adult age card and (3) Australian identity card may make identity management processes easier for clients and more cost efficient for government and private sector organisations.

While currently there is no proposal to remove any of the existing cards or other products from the Australian market, it is expected that one of the three primary ID cards would be used to first establish or enrol an individual's identity to an organisation which may then decide to issue their existing organisation's card or other identity product to the person if they wish.

The proposed use of smartcard technology in the proof of identification process will permit access to a client's personal data held by transport authorities to be system controlled. This practice will in turn strengthen both client privacy and the security of client data.

The use of smartcard technology that is to be contained in the three primary identity products is also expected to streamline, through automation, and speed up the verification of a person's identity. The technology also promotes the opportunity for clients and organisations to choose to store other information in the smart chip of these three products.

Governments might provide some applications and associated data on the smart chip as a mandatory requirement. This may mean that some existing government products such as a Medicare card, a gun license and/or a marine license may no longer be issued separately. However, clients should also be able to choose to have other non-government products (including club memberships, loyalty programs, applications that manage access to private networks or even manage door access to secure areas) added to any of their three core identity products.

It should be noted that access to the smart chip by an organisation would be government controlled. An organisation would need government approval to access smartcard products and load applications/data onto the smart chip.

The provision of an Australian identity card should be available to Australian citizens *as an option*. At the time of writing this book, demand for the Australian identity card is not anticipated to be high. It is expected that most Australians will continue to use their driver's licence as their main proof of identity document. However, those adult Australians (such as the elderly) who do not have a driver's licence and, instead, use an adult age card or other product as their proof of identity document may consider the Australian identity card to be a more appropriate personal identity document.

It is quite likely that privacy and civil liberty groups as well as the media may comment strongly on any government attempt to strengthen identity management within the Australian community. Some groups may continue to draw comparisons with the Hawke era Australia card proposal and possibly resurrect a Big Brother scare campaign.

However, there are now legislative requirements and processes in place to make Australian governments and the private sector more accountable and better able to manage personal identity than may have been available in the mid-80s. These days, the community as a whole is better educated about identity management and the value of their identity and hence there is greater awareness of the value of strong identity security, including the benefits of a national identity card.

The community may also better appreciate the efforts by government to protect their identity, improve cyber security and reduce the high costs of identity crime within the community.

A sign of the overall success of this proposal may be the scenario whereby most Australians provide tacit support to the proposed improvements to personal identity management but dismiss them as having no impact on their own situation. According to this proposal, most Australians should not be required to take any action or have to change what they are currently doing to prove their identity.

CONCLUSION

Traditionally, Australian state and territory transport authorities have been the recognised providers of the community's key domestic identity product, which is the driver's license. The driver's license is provided for the primary purpose of road safety. In the absence of a better domestic product, the community has adopted the driver's license as its main personal identity document.

The state and territory transport authorities' current role in the management of personal identity has been further cemented by the provision of adult age cards to the community. These cards are not connected with road safety, but have a strong link to personal identity management.

Over the years, transport authorities have transitioned from their primary role of maintaining road safety to championing the defence of the community against identity crime. It could be argued that transport authorities have periodically updated the driver's license product to ensure it is more resistant to fraud, not purely for the purpose of road safety but more from the perspective of reducing identity crime.

The Australian community is currently experiencing increasing levels of threat involving identity crime that must be addressed. The present costs and problems associated with identity crime to the Australian community are high.

Moreover, this issue is continuing to escalate and is becoming increasingly difficult to address. Identity crime is occurring in face-to-face situations but also in cyber environments. Australian citizens and organisations are incurring high financial losses as a direct result of cybercriminals. At present, there seems to be little evidence of an adequate response to cybercrime by government.

Identity crime and the ability to respond adequately to these crimes through improved identity management is both a national and international interest. However, improvements to identity management within any community can only be government led.

Internationally, a new solution is required to managing personal identity within the cyber environment – in particular, to reduce anti-social and criminal activities. These activities may range from cyberbullying, communication between organised criminal gangs and other entities, financial fraud and coordination of terrorist strikes against the community.

The Austroads organisation created a forum and a common direction for transport authorities which enabled them to unite and provide a critical defence of the community against identity crime. The Australian Federal Government should be commended for projecting the issue of identity crime in the mid-2000s into the political arena. It is promising to observe that the federal government has attempted to position itself in a national coordination role in terms of personal identity management, but the reality is that they cannot readily control this burgeoning problem on the domestic front. For effective management of personal identity, the federal government must rely on the states and territories.

At present, the Queensland Department of Transport and Main Roads is responsible for providing the most secure domestic identity products (which are the smartcard driver's license and adult age card products) available to the community. The Queensland products are highly resistant to fraud; however, the smartcard features of these products are currently under-utilised.

Australian governments are currently working to strengthen processes associated with identity management within the face-to-face environment. While it is no doubt a point of discussion within government circles, the most successful (and straightforward) action would be to improve the driver's license and adult age card products so that they are more resistant to fraud. This could be achieved by immediately updating all Australian drivers' licenses and adult age cards to smartcard products so that they are consistent with the Queensland products.

Of concern is failure by governments to protect the community from crimes and anti-social activities in the cyber environment. This is not a road safety issue and so not the role of transport authorities, but it is the role of government. It is proposed that the Australian Federal Government should assume a leadership role in terms of personal identity management for its citizens.

An ideal outcome for this issue would be to have a strong connection between an electronic identity and the actual person. This could be achieved by adding an electronic identity to an individual's main personal identity document which in Australia is a person's driver's license, adult age card or Australian identity card – if it becomes available. This process involves leveraging off the existing dominant identity product and the traditional managers of identity within the community. These identity cards should become the vehicle for issuing the community with electronic identities.

Smartcard technology enables the provision of electronic identities to the community be means of the driver's license product. Adding an electronic identity to the smartcard driver license product would also allow state and territory transport departments to continue to provide operational management of identity to the Australian community.

Electronic identities should be available to support existing anonymous transactions, such as entry into an entertainment venue. However, an electronic identity should also be available to provide secure online transacting between a citizen and private or government organisations, potentially using processes such as Public Key Infrastructure.

The addition of smartcard technology to the three core identity products may also provide the opportunity to replace some existing physical identity products issued by government and private sector organisations. This should occur where an individual or a government organisation wishes to have an electronic application and data reside on the smart chip of an identity product.

Generally speaking, Australia is not lagging behind the rest of the world in terms of its identity management practices, but it certainly is not a leader in this arena. Strategically, there does not seem to be any evidence that the government is preparing to improve identity management to address current problems with cybercrime or to prepare the community for a future that is more reliant on electronic communications and transacting.

Australia certainly seems to be behind the European Union which is preparing to provide its citizens with an electronic identity that supports online transacting. Singapore and United States are also conducting research in preparation to be able to offer their citizens with the capacity to transition to a secure online business environment.

Government and private sector organisations should be preparing for a future which involves 24-hour service delivery that is provided at low cost and high speed. All services (including high risk services) should be transitioned to an electronic environment and clients should not be required to attend a face-to-face service outlet. This process cannot occur without significant improvements to existing identity products and the provision of a highly trusted electronic identity.

In comparison to other countries, Australia is well-placed to quickly improve its identity management practices. But government must be prepared to modernise its strategies to better position Australia to meet existing and future challenges in identity management. Protecting the community from crime (including identity crime) is a key responsibility of government. However, this objective should be achieved while also actively embracing any new opportunities that may allow the community to further advance and prosper.

GLOSSARY

Below is a list of acronyms, initialisms and terms that have been used in this book.

Term	Explanation
AAMVA	American Association of Motor Vehicle Administrators. This organisation promotes consistency amongst transport authorities who issue drivers' licenses in the United States and Canada.
Access Card	A proposed federal government identity card to be used by the recipients of social service payments. The card was proposed in the mid-2000s but was never introduced.
Adult Age Card	A government-issued proof of identity card designed to prove an individual is over 18 years of age and can enter an establishment that serves alcohol.
Australia Card	A proposed federal government identity card to be issued to all Australians. The card was proposed in the mid-1980s but was never introduced.

Australian Identity Card	A proposed government-issued proof of identity card to be developed and to be made available to anyone who meets the criteria set by the Australian Federal Government.
Austroads	A national organisation that is tasked with supporting national coordination of state transport authorities.
Big Brother Concept	A concept that first appeared in George Orwell's book *1984*. The concept centres on the idea that government protects the community through constant surveillance.
Biometric technology	A range of technologies that capture and measure the unique characteristics of a person's particular biometric, such as their fingerprints.
Birth certificate	An official name document issued by the state and territory Office of Births, Deaths and Marriages
COAG	Council of Australian Governments. Committee that contains the federal government and state and territory governments. The meeting is usually attended by the prime minister and premiers.
Data encryption	Converts data into a code. It is used to make data secure from unauthorised access.

Digital image	An image captured with a digital camera.
DVS	Document Validation Service. A national service to improve the validation of a range of personal identity documents.
E-Gate Card	A passport control identity card used in the United Arab Emirates to enter and leave the country without interaction with immigration officers.
Electronic identity	An identity provided to a person that allows them to authenticate themselves (typically in an online transaction).
EU	European Union
Federal Government	Australian National or Federal Government
GCC	Gulf Cooperative Council. The Council contains six countries of the Gulf region who often function as single block of countries in international affairs.
Gun Licence	A state-issued license that permits a person to possess a registered firearm.
Identity crime	A generic term used for crimes in which an individual has created a fake identity or stolen another person's identity.
Identity theft	An individual has stolen another person's identity and used that identity to commit a crime.

Image database	A computer system that is designed to store and manage digital facial images of clients.
Medicare Card	The Australia Federal Government card issued to most Australians to allow them to access healthcare services.
NEVDIS	National Exchange of Vehicles and Driver Information System. A national system that facilitates the exchange of driver license and vehicle registration information between Australian transport authorities and police.
NISS	National Identity Security Strategy. An agreement between the federal government and state and territory governments to achieve better management and protection of personal identity.
OBDM	Office of Births, Deaths and Marriages
PIN	Personal Identity Number
Queensland Transport	Queensland Department of Transport and Main Roads
Service Delivery Network	A network across Australia that provides in-person service for driver license products. It also includes telephone and Internet service delivery options.
Singpass	An online gateway to government services provided by the Singaporean government for its citizens.

Smartcard	A plastic card that contains a smart chip or minicomputer that is embedded in the plastic.
Smart chip	A minicomputer that is embedded in a plastic card. It may contain a contact or a contactless process for reading data from the computer.
Transport Authority	A state or territory organisation which is responsible for issuing drivers' licenses.
UAE	Country of the United Arab Emirates
US	Country of the United States of America

BIBLIOGRAPHY

PART ONE

Australian Attorney-General, *Identity Crime and Misuse in Australia 2016.* https://www.ag.gov.au

Australian credit card and debit card statistics 2017. https://www.finder.com.au

Austroads, *Home.* http://www.austroads.com.au

Austroads, *About NEVDIS.* http://www.austroads.com.au

Bank NAB Brochure. *100 Points guide – all you need to open an account*

Commonwealth of Australia, *National Identity Security Strategy.* https://www.ag.gov.au

Hannula, Lauren, *The Real ID Act: Are You Ready for a National ID?* http://www.dmv.org

IT news, *Unisys scores five-year Qld digital driver's license deal.* http://www.itnews.com.au

Lopez, Antonio. *Staring Privacy in the Face.* http://besser.tsoa.nyu.edu

Science, *How biometrics works.* http://science.howstuffworks.com

TC, Inside India's Aadhar, *The World's Biggest Biometrics Database*. https://techcrunch.com/2013/12/06

Wikipedia, *100 Point check*. https://en.wikipedia.org

Wikipedia, *2014 Sydney Hostage Crisis*. https://en.wikipedia.org

Wikipedia, *Yahoo Data Breach*. https://en.wikipedia.org

PART THREE

ABC News, *Canberra has become a parallel moral universe*. http://www.mamamia.com.au/australian-politicians-benefits/

Australian Attorney-General, *Document Validation Service*. https://www.ag.gov.au

Australian Attorney-General, *National Identity Proofing Guidelines 2016*. https://www.ag.gov.au

Australian Attorney General, *Protecting and Recovering Your Identity*. https://www.ag.gov.au

Austroads, *Registration and Licensing*. http://www.austroads.com.au

Department of Attorney General, W.A., *Births, Deaths and Marriages*. http://www.bdm.dotag.wa.gov.au

News.com.au, *Aussies panicking over Centrelink demands to pay up to avoid debt collector*. http://www.news.com.au

Office of the Australian information Commissioner, Privacy Fact Sheet 17 – *Australian Privacy Principles*. https://www.oaic.gov.au

Queensland Government, *Reasons for License Suspension*. https://www.qld.gov.au.

Queensland Government, *Renewing Your License Online*. https://www.qld.gov.au.

Queensland Government, *Transport Operations (Road Use Management—Driver Licensing) Regulation 2010, Regulation 133.*

Regulation (EU) No 910/2014 of the European Parliament and of the Council of 23 July 2014, *on electronic identification and trust services for electronic transactions in the internal market and repealing directive 1999/93/EC.* http://eur-lex.europa.eu

PART FOUR

EFA, Access Card/National ID Card. https://www.efa.org.au
Wikipedia, *Australia Card.* https://en.wikipedia.org
Wikipedia, *Big Brother (1984).* https://en.wikipedia.org

PART FIVE

ABC news, *Electronic pickpocketing' looms as next threat in credit card fraud, police, security experts say.* http://www.abc.net.au/news/2014-05-30
Australian criminal intelligence commission, *Illicit firearms in Australia.* https://www.acic.gov.au
SBS news, *How easy is it to get a gun in Australia?* http://www.sbs.com.au/news/article/2014/12/18
Nine News, *Pauline Hanson calling for national ID card.* http://www.9news.com.au
Seven News, *Germany threatens online giants with 50 mn euro hate speech fines.* https://au.news.yahoo.com
The Age Newspaper, *Tap and Go Credit Card Fraud, Chewing up Police Resources.* http://www.theage.com.au
The Guardian, *Queensland lockout laws set to pass after deal with Katter party MPs.* https://www.theguardian.com/australia-news/2016/feb/17
Wikipedia, *Smart Cards.* https://en.wikipedia.org

PART SIX

AAMVA, EID Working Group, *Identification Security*. http://www.aamva.org

AAMVA, Real ID, *Identification Security*. http://www.aamva

Computerworld, *UK Government Identity Scheme – GOV.UK*. http://www.computerworlduk.com

Department of Border and Immigration, *Fact Sheet – New Zealanders in Australia*. https://www.border.gov.au

Dubai FAQ is guide to Dubai, *Egate cards in Dubai and UAE*. http://www.dubaifaqs.com

Dubai Roads and Transport Authority, *Document Validation Service*. https://traffic.rta.ae

European Commission, E-Identification. https://ec.europa.eu

European Union Act (CAP. 460), *Identity Card and other Identity Documents Act Order, 2012*. http://justiceservices.gov.mt

Hannula,Lauren. The Real ID Act: Are you ready for a national ID. http://www.dmv.org

Information Age, *UK government's Identity Management Plan*. http://www.information-age.com
Khaleej Times, *Emirate's ID card to be mandatory*. http://www.khaleejtimes.com

Regulation (EU) No 910/2014 of the European Parliament and of the Council of 23 July 2014, *on electronic identification and trust services for electronic transactions in the internal market and repealing directive 1999/93/EC*. http://eur-lex.europa.eu

Smith, Oliver. *Dubai airport the world's busiest airport for international passengers in 2016*. http://www.traveller.com.au

Straits Times, *And next, a mobile digital ID.*
http://www.straitstimes.com
Straits Times, *Plans for all Singaporeans to get new e-IC for online transactions on the cards.*
http://www.straitstimes.com/tech/and-next-a-mobile-digital-id

UK Government Home Office, *Identity cards and new Identity and Passport Service suppliers.* https://www.gov.uk

Wikipedia, *National Registration Identity Card.*
https://en.wikipedia.org

Wikipedia, *Driver's Licenses in United States.*
https://en.wikipedia.org

ABOUT THE AUTHOR

Steve Venning was employed for over 30 years by the Queensland government, Australia. For a significant part of that time he was involved in the driver licensing business. In addition, he was Queensland's representative on the Austroads subcommittee, focusing on national consistency of driver licence and vehicle registration services.

During his employment with the Queensland government, Steve also completed a Bachelor of Arts degree, a Master's in Business Administration and a Master's in Information Systems from the University of Queensland and the University of Central Queensland (Australia).

From 2008 to 2014 he was employed as a business consultant by the Government of Dubai, United Arab Emirates. In this role, Steve assisted in modernising government business operations dealing with driver licensing and vehicle registration services.

www.ingramcontent.com/pod-product-compliance
Lightning Source LLC
Chambersburg PA
CBHW072141270326
41931CB00010B/1834